Of Knyghthode and Bataile

Middle English Texts Series

The Middle English Texts Series produces scholarly texts designed for research and classroom use. Its goal is to make available to teachers, scholars, and students texts that occupy an important place in the literary and cultural canon but have not been readily available in print or online editions. The series does not include authors, such as Chaucer, Langland, or Malory, whose English works are normally in print. The focus is, instread, upon Middle English literature adjacent to those authors that are needed for research or teaching. The editions maintain the linguistic integrity of the original work but within the parameters of modern reading conventions.

Of Knyghthode and Bataile

Edited by
Trevor Russell Smith and Michael Livingston

A publication of the Rossell Hope Robbins Library,
the University of Rochester,
and the Teaching Association for Medieval Studies (TEAMS)

by

Medieval Institute Publications
Kalamazoo, Michigan
2021

The text of this book was set in Arno Pro 11pt, a contemporary typeface inspired by books from .the 15th and 16th centuries. It was designed by Robert Slimbach at Adobe

Library of Congress Cataloging-in-Publication Data

Names: Vegetius Renatus, Flavius. Epitoma rei militaris. | Smith, Trevor Russell, 1985- editor. | Livingston, Michael, 1975- editor.
Title: Of knyghthode and bataile / edited by Trevor Russell Smith and Michael Livingston.
Description: Kalamazoo, Michigan : Medieval Institute Publications, 2021. | Series: Middle English texts series | Includes bibliographical references. | Summary: "Composed for King Henry VI in the middle of the Wars of the Roses, Of Knyghthode and Bataile adapts the most widely used military manual in the Middle Ages into English verse. Responding to both the evolution of warfare and the historical background of his own time, its anonymous poet produced what one critic has called "one of the most brilliant military poems of the fifteenth century." That work is here re-edited from all four surviving manuscripts, and presented with a contextualizing introduction and copious notes and glosses"-- Provided by publisher.
Identifiers: LCCN 2021008678 (print) | LCCN 2021008679 (ebook) | ISBN 9781580444767 (paperback) | ISBN 9781580444750 (hardback) | ISBN 9781580444774 (pdf)
Subjects: LCSH: Military art and science--Early works to 1800. | Knights and knighthood--England--History--To 1500. | Military art and science--Poetry. | Didactic poetry, English (Middle)
Classification: LCC U101 .O365 2021 (print) | LCC U101 (ebook) | DDC 355--dc23
LC record available at https://lccn.loc.gov/2021008678
LC ebook record available at https://lccn.loc.gov/2021008679

ISBN 9781580444767 (paperback)
ISBN 9781580444750 (hardback)

Printed and bound by CPI Group (UK) Ltd, Croydon, CR0 4YY

Table of Contents

Acknowledgments

Our gratitude, first and foremost, to Russell A. Peck, General Editor of the Middle English Texts Series for allowing us to add this fascinating poem to its ever-growing list of books. Assistant Editor Pamela Yee has been extraordinarily helpful in moving the volume through its various stages of production. Staff editors Mead Bowen and Ashley Conklin were instrumental in completing the most labor-intensive steps of production, the manuscript read and bibliography checks. Alan Lupack and Anna Siebach-Larsen joined them in providing an extradornary number of helpful suggestions to improve the text.

We also owe thanks to James Freeman at Cambridge University Library for examining the collation of the Cambridge manuscript and also for checking some of its difficult to read bits under ultraviolet light, to Hanno Wijsman at the Institut de recherche et d'histoire des textes, Paris, for assisting with the Sankt-Peterburg manuscript, and to the many librarians and keepers of manuscripts for their helpfulness, expertise, and patience.

Introduction

Of Knyghthode and Bataile is an unexpectedly fascinating work.[1] At one level, many scholars have no doubt dismissed it as a point of trivia: it is the second surviving English rendition of one of the most popular military treatises ever written. Yet in truth it is much more than that. "One of the most brilliant military poems of the fifteenth century," as Catherine Nall has regarded it, *Of Knyghthode and Bataile* well deserves to be counted among "the most elaborate, creative, and rich texts produced during the Wars of the Roses."[2]

This poem, as Nall indicates, stands at the crossroads of history. Behind and beyond its status as a translation — more accurately, a paraphrase — of Vegetius's famed work alternatively titled *De re militari* or *Epitoma rei militaris*, it engages with the contemporary realities of the nascent Wars of the Roses, the evolving status of chivalry at the end of the Middle Ages, and the shifting face of war as technological changes brought gunpowder to conflicts on the land and great ships into conflict upon the sea. The poet's world, in so many senses, was fragmenting, and he saw in the historical Vegetius a means to achieve present unity and healing: a properly trained fighting force would enforce the legitimate authority of the king, and a stabilized throne would bring peace for a shared society.

First Years of the Wars of the Roses

In its most literal sense, the Wars of the Roses began on May 22, 1455, when longstanding political rivalries among noble factions, combined with the mental instability of Henry VI, led Richard, duke of York, to meet the king in arms at the town of St. Albans. What would come to be called the First Battle of St. Albans had an impact far beyond its relatively small scale: Lancastrian leaders, including the duke of Somerset, the earl of Northumberland, and Lord Clifford were killed in the Yorkist victory, and the king was abandoned into his enemies' hands. The duke of York made no immediate play for the throne — whether due to respect for regnal authority or an awareness of the limitations in his position — but in retrospect the die of war had been cast. The Wars of the Roses, which would last until the 1487 death of the duke's son, King Richard III at Bosworth Field, had begun.

In the early years, attempts at keeping the peace were made, most notably during the elaborate ceremonies of the Loveday of 1458. On March 25 of that year, Henry VI — at the time in complete control of his faculties — orchestrated a public display of peace and unity, walking with his Yorkist foes from Westminster Abbey to St. Paul's Cathedral in London, surrounded by pomp, pageantry, and armed retinues. His queen, the influential Margaret of Anjou who fought bitterly for her husband's Lancastrian cause, followed the king, walking hand-in-hand with the duke of York. Payments and promises were made.

---

1 The previous edition of the poem (see below), and thus the driver of critical conversations about it, entitled the work *Knyghthode and Bataile,* but both the text itself and its earliest surviving witness are consistent in naming it *Of Knyghthode and Bataile* — a nod, no doubt, to its Latin roots.

2 Nall, *Reading and War*, pp. 114, 138.

Within months, however, hostilities were once more rising. The Yorkist Richard Neville, the earl of Warwick whom history would come to call "the Kingmaker," had held the captaincy of the garrison at Calais for several years. Calais, a massive port on the shores of France, was not just a rich nexus of trade between the English, the French, and the Low Countries, but also an important foothold in the larger geopolitical struggles between the kingdoms of France and England. To be captain of Calais was an important and enriching post. Beginning in May 1458, Warwick directed ships from Calais to plunder a number of Castilian and Hanseatic merchant ships, causing a diplomatic row. Called by Henry VI to answer charges on the matter, Warwick declined.

Influential members of the king's retinue, led by Margaret of Anjou, assumed the worst of Warwick's actions. The court retreated to Coventry, deep in the queen's home turf, and a council was called for June 24, 1459. Fearing that attendance would mean arrest, the duke of York, the earl of Salisbury, and the earl of Warwick refused the summons. Branded as rebels, the three men set out to bring their scattered forces together under the banner of York at Ludlow. On October 12 the rebels were making lines near a bridge beside the small town of Ludford in Shropshire, when the royal banner appeared on the horizon. The rebel army quaked at the prospect of fighting the king himself, and the Yorkist leaders knew they were undone. With hardly a shot fired, the Battle of Ludford Bridge was over before it began. York himself fled to Ireland, where he still had support. Salisbury and Warwick fled to Wales and then to Calais, only just beating the arrival of the duke of Somerset, whom the king had ordered to replace the rebellious earl as captain of Calais.

From November 20 to December 20, the Parliament — later termed the Parliament of Devils — met in Coventry. The Yorkist rebels were declared guilty of high treason, and bills of attainder were passed against them. Their lands were seized. Henry VI and the Lancastrians began to wrest back control of the country, all the while keeping a watchful eye across the English Channel to where Warwick adeptly held off Somerset's attempt to take back Calais and entrenched his power.

Date and Authorship

The following spring, a parson exiled from Calais in the political struggle approached Lord Beaumont with the gift of a poem about the making of war that he wished to give to the king. As Nall has noted, this was in keeping with the times: "reading, writing and the prosecution of warfare went hand in hand in the fifteenth century."[3] In any case, the parson's poem was read, found worthy, and the presentation was made when Henry VI returned to London on March 1, 1460. The poem was *Of Knyghthode and Bataile,* and the parson seems likely to have been a churchman named Robert Parker.

This identification of the date and authorship of *Of Knyghthode and Bataile* stands both with and against prior scholarship. The poem's first and only edition was executed for the Early English Texts Society in 1935 by Roman Dyboski and Zygfryd Marjan Arend, based on the three manuscripts then known: Cambridge, Pembroke College, MS 243; London, British Library, MS Cotton Titus A.xxiii; and Oxford, Bodleian Library, MS Ashmole 45. A fourth manuscript has recently been discovered in London, College of Arms, MS R.25.[4] In their edition, Dyboski and Arend are led by the poem's opening stanza — which declares its occasion to be a festive entrance by the king into London on the kalends of March (lines 1–8) — to suggest that the poem ought to be associated with the Loveday of 1458. This event was indeed celebratory, but Dyboski and Arend are forced into substantial difficulty trying to turn March 25 (the date of the

[3] Nall, *Reading and War,* p. 10.

[4] Davies and Edwards, "New Manuscript of *Knyghthode and Bataile.*"

Loveday) into March 1 (the poem's kalends of March).[5] Even more problematic, the poem makes reference to events beyond the Loveday. Lines 985–1026 unmistakably refer to the Battle of Ludford Bridge on October 12, 1459 and its immediate aftermath: Warwick's subsequent retreat to Calais on November 2, 1459 (lines 987–88), and the Coventry Parliament from November 20 to December 20, 1459 (line 995). This Parliament's bills of attainder against the Yorkists also lies behind other statements on the part of the poet, including his direct reference to the king's enemies as "a legioun attaynte" (line 2017).

In short, the poem must surely have been composed *after* the end of 1459. As for a date *before which* it must have been written, the partisan poet would no doubt have crowed about the great Lancastrian victory at the Battle of Wakefield on December 30, 1460 if he had known of it. Narrowing the range still further, it would be difficult for the poet to present the poem through the intermediary of Lord Beaumont after the Battle of Northampton, on July 10, 1460, in which Beaumont died. A presentation date of March 1 not only falls perfectly within these terminal dates but also fits the approximate occasion of King Henry VI's return to London from Coventry. And, as Daniel Wakelin has observed in likewise arguing for a 1460 dating, "the scanty records of Saturday, 1 March 1460 do not attest a full entry pageant in London, but they do reveal martial display surrounding the king" that could inspire the vision described by the poem.[6]

Regarding the parson in question, Dyboski and Arend conclude that "the material at our command proves insufficient for identifying the person of the author."[7] Other scholars have not been so reserved. Earlier, in 1913, Henry Noble MacCracken made the suggestion that the poet ought to be identified with Robert Parker, whom he regarded as the author of another anonymous work, *On Husbondrie,* likewise a verse adaptation of a late-antique Latin treatise.[8] Parker's career, MacCracken observed, had much to recommend it as would seem fitting for the author of *Knyghthode and Bataile.* The Patent Rolls note that a Robert Parker, chaplain, succeeded a clerk of the king's closet to an appointment as parson of Stanford Rivers on February 25, 1439; it is presumably this same Robert Parker, now declared the king's own clerk, who was later named parson of the Church of St. Nicholas in Calais on August 16, 1450.[9] Robert Parker also appears on March 2, 1460 — the day after what we can now identify as the presentation date of the poem[10] — in a military-related commission from the king:

> Commission to John Judde, esquire, master of the king's ordinance, Henry Nevill, Alexander Norton, Robert Parker, John Carpenter and Dederic Tyle [*rectius* Pyle], to take carpenters called 'whelers' and 'cartwryghtz' and other carpenters, stonemasons, smiths, plumbers, artificers and workmen for the works of the king's ordnance, and bombards, cannons, 'culvryns,' 'serpentyns,' crossbows, bows, arrows, 'saltpetre,' powder for cannons, lead, iron and all other stuff for the said ordnance, and carriage therefor and horses called 'hakeneys.'[11]

[5] Ed. Dyboski and Arend, *Knyghthode and Bataile,* pp. xvi–xvii.

[6] Wakelin, "Occasion, Author, and Readers," p. 263. For an alternate reading of the poem's depiction of the entry, which focuses less on its historical reality and more on its literary effects, see Scase, "Writing and the 'Poetics of Spectacle'," pp. 181–82.

[7] Ed. Dyboski and Arend, *Knyghthode and Bataile,* p. xxiv.

[8] MacCracken, "Vegetius in English," pp. 398–400.

[9] *Calendar of Patent Rolls* 3.242, 5.334.

[10] MacCracken, as Dyboski and Arend would later do, connects the poem with the Loveday of 1358 ("Vegetius in English," p. 394).

[11] *Calendar of Patent Rolls* 6.605.

Despite the suggestive possibilities of this passage, to which we will return, MacCracken's identification of Robert Parker as the author of *Knyghthode and Bataile* was rejected in 2004 by Wakelin, who put forward John Neele instead.

Wakelin begins his objection to Robert Parker by first observing several flaws in identifying shared authorship between *On Husbondrie* and *Of Knyghthode and Bataile.*[12] These points are rightly made, though they say nothing of Robert Parker's hand in any case, just that the author of one might not be the author of the other. And, as it happens, the case for Parker's presumed authorship of either text lies most strongly with *Of Knyghthode and Bataile*, not *On Husbondrie.* Wakelin next wonders whether Beaumont's involvement in the presentation of the poem might indicate that the poet has a place in the court of Margaret of Anjou, for whom Beaumont served as her "grandest household servant," though this train of speculation does not go far.[13] Returning to Parker's candidacy, Wakelin states that "it is unclear how long Parker remained at St Nicholas's, Calais: other priests were appointed to the chantry of Holy Cross in that church throughout the 1450s." In addition, he dismisses the possibility that the parson Robert Parker is the same Robert Parker commissioned by the king to deal with armaments in 1460: "another unspecified Parker was employed as armourer to Henry VI in 1455, who is surely *that* Parker."[14]

Wakelin is quite right that there was a Parker who served as king's armorer at this time, but that man is not actually "unspecified" in the records. His name was Thomas, and he was given a grant for life "of all the workshops . . . of the armoury within the Tower of London" on May 6, 1450; he also later served as a counsel to a trial by battle on May 11, 1453.[15] It is a difficult proposition to accept that the keeper of the king's accounts was on multiple instances confused by Robert and Thomas Parker in the absence of any evidence.

In place of Robert Parker, Wakelin suggests John Neele, who, "because of his Lancastrian affiliations and because of his learning," was appointed rector of St. Mary's in Calais in January 1458.[16] To the latter point, Neele's education cannot serve as much argument for or against his authorship without a direct connection between his schooling and the text in question or, at the very least, evidence that Parker's education would have prevented him from composing it, none of which is evident. To the former point, Neele holds no more Lancastrian affiliations than Parker: Neele was indeed the receiver of several grants from the king in 1460; but so, it seems, was Parker, who had previously served as a clerk of the king. It is true that in May and June 1460 this same Neele apparently "received a benefice on Guernsey and a command to set in order castles there and in Jersey" — evidence that Wakelin posits as a reward for composing *Of Knyghthode and Bataile.*[17] Yet, as we have already seen, Parker appears to have a military command of his own, a commission to oversee the manufacture of weapons of war for Henry VI on March 2, 1460, just *one day* after the supposed presentation of this poem about war to the king.

Wakelin's last piece of evidence in favor of Neele is that his later career matches the poem's later history: Neele continued to be held in favor by the crown after Edward IV seized it, just as later Yorkist manuscripts preserved *Of Knyghthode and Bataile* by stripping or altering its specifically Lancastrian segments as they-copied it.[18] This logic seems to suggest authorial oversight of those alterations, though the manuscripts

12 Wakelin, "Occasion, Author, and Readers," p. 261.

13 Wakelin, "Occasion, Author, and Readers," pp. 263–64.

14 Wakelin, "Occasion, Author, and Readers," p. 264.

15 *Calendar of Patent Rolls* 5.314; Ed. Nicolas, *Proceedings and Ordinances*, 6.129–30.

16 Wakelin, "Occasion, Author, and Readers," p. 265.

17 Wakelin, "Occasion, Author, and Readers," p. 265.

18 Wakelin, "Occasion, Author, and Readers," p. 266. On these alterations, see pp. 11–14, below.

provide no internal evidence that this is so. And, once more, we can say much the same about the career of Parker, who was parson of St. Gregory by St. Paul's during the reign of Edward IV.[19] Wakelin suggests that this cannot be the same Robert Parker because this would give him an over-long career. Yet we can be sure that at least one well-connected churchman named Robert Parker had an overlapping career across this entire period: a clerk named Robert Parker, son of John Parker, is mentioned in a deed of January 17, 1434, and a chaplain named Robert Parker, son of John Parker, is recorded on December 1, 1487.[20] We cannot be certain that this is the same Robert Parker as the chaplain in Calais and the man commissioned by the king, but, likewise, it should not be casually dismissed. Lastly, it is worth noting, too, that the poet's devotion to Calais comes across as long-held, with deep enmity for the Yorkists who, in the text, now hold it. Parker was assigned to a Calais post in 1450; shortly afterward, the duke of Somerset — one of the Lancastrian leaders who was killed in the First Battle of St. Albans — became captain of Calais. In the political struggle of the next decade, Parker would have seen the captaincy come into the hands of the duke of York himself in 1454, and then the Yorkist earl of Warwick in 1456. Neele, who was appointed to his Calais post in 1458, would have known Calais only as a Yorkist stronghold, while Parker would have personally witnessed, as *Of Knyghthode and Bataile* relates, its fall from the Lancastrians.

Beyond the poet's self-identification with Calais, *Of Knyghthode and Bataile* gives further reason to connect poem and place. In several key sequences, the poet "shows great familiarity with and takes a wild sort of delight in stormy aspects of the sea," as Dyboski and Arend observe, suggesting that "he must many a time have observed them from his town of Calais."[21] The poet's imaginative explication of a naval battle in the latter parts of the poem — a sequence that spurred MacCracken to comment "Here is someone, in that barren age, who knows what he is about"[22] — likewise seems to point to a life lived in close proximity to the sea. Unfortunately, not enough is known about our possible authors to utilize this awareness to help us identify more positively the poet at work.

On balance, the identification of the author cannot be made with complete certainty, but there is little to Neele's claim that Parker cannot match or better.[23] One must admit that it is a highly remarkable coincidence that Robert Parker was close to the king, was a parson of Calais prior to its Yorkist takeover, and was named to an appropriate commission within a day of his formal presentation of the poem. It is also noteworthy to observe that in that commission Parker and his fellows were specifically tasked with ordinance including bombards, cannons, culverins, and serpentines. These relatively new-for-the-time artillery pieces make appearances in two memorable sequences in *Of Knyghthode and Bataile*:

> Al this aray, and bumbardys thei cary, *bombards*
> And gunne and serpentyn that wil not vary, *[a] gun and serpentine*

---

19 The National Archives, SC 1/46/265.

20 Wiltshire and Swindon Archives, 212B/4796; Devon Record Office, 3248A–0/11/117. If these are the same Robert Parker who served the king in Calais and London, then the first record seems to preclude the possibility that Thomas Parker in the king's armory was Robert Parker's immediate relation: it lists Robert's brother as John Parker, son of John Parker.

21 Ed. Dyboski and Arend, *Knyghthode and Bataile*, p. xxxvi.

22 MacCracken, "Vegetius in English," p. 395.

23 That is, unless line 2982 could definitively be traced to John's name, on which see pp. 16–18, below.

Fouler, covey, crappaude, and colveryne[24]
And other soortis moo then VIII or IX^{ne}.

. . . .

The canonys, the bumbard, and the gunne, — *cannons; bombard; gun*
Thei bloweth out the voys and stonys grete, — *sound and large stones*
Thorgh maste and side and other be thei runne. — *mast; hull*
In goth the serpentyne aftir his mete. — *serpentine; target*
The colveryne is besy forto gete — *coulovrine*
An hole into the top. And the crappaude — *top [of the ship]; crapaudeau*
Wil in. The fouler eek wil have his laude. — *veuglaire also; praise*
(lines 1849–52, 2854–60)

Indeed, these citations, and the royal commission that came the day after its presentation, are some of the earliest citations of several of these gunpowder weapons in English.

So what was the text that Robert Parker — or John Neele, or perhaps someone else entirely — handed over to King Henry VI?

Vegetius

Sometime in the late fourth century or early fifth century, a Christian writer named Publius Flavius Vegetius Renatus — commonly and hereafter called Vegetius — wrote a treatise on how to reform the declining Roman army into the more potent force that it had been in earlier days.[25] Scholars continue to debate the degree to which his work, *De re militari* [*Of Military Matters*], accurately reflects the Roman military experience in either his own or an earlier time, especially considering its heavy dependence on earlier literature, yet what is beyond dispute is the enormity of its influence in later centuries.[26] Vegetius's book is the most influential military treatise across the entirety of the Middle Ages: it was copied and recopied throughout the period and survives in nearly two-hundred manuscripts in its original Latin version alone.[27]

Through translation and paraphrase, Vegetius found additional life in the vulgar languages of Europe, passing through the hands of writers as well known as Jean de Meun, Christine de Pizan, and Niccolò Machiavelli.[28] The first translation into English was a 1408 prose translation, apparently by John Walton. Among its eleven surviving manuscripts is an ornate copy made for King Richard III.[29]

The second Vegetius in English is the work here edited, in which our parson from Calais was determined to adapt this important text into Middle English poetry for King Henry VI. For its early date and historical context alone, it is noteworthy. As Julia Boffey writes: "whatever its origins, the rendering into English of a treatise about 'euery feat of werre' by a one-time inhabitant of a town whose role throughout

---

24 *Veuglaire, covey, crapaudeau, and culverin*

25 On the dating of Vegetius, see Goffart, "Date and Purpose" and Charles, *Vegetius in Context.*

26 Sherwood, "Studies in Medieval Uses," pp. 39–45.

27 So well-known was Vegetius, in fact, that the author's name became a short-hand reference to military writings whether of his hand or not, akin to calling a generic tissue a "Kleenex," due to the ubiquity of the brand name. Thus when Gower references "the clerk Vegecius" as his source for the tale of Ylia (*Confessio Amantis*, ed. Peck, 5.885), he could be referring to a number of related texts.

28 Allmand, *The "De Re Militari" of Vegetius*, pp. 156–59, 121–27, 139–47.

29 London, British Library, Royal MS 18.A.xii.

the fifteenth century was crucial in hostilities relating to trade and to both domestic and foreign politics is not without its significance."[30]

Despite containing many words and turns of phrase that bear French influence, *Of Knyghthode and Bataile* appears to be adapted from a particular family of the Latin Vegetius.[31] A thorough comparison of its peculiarities with the Middle English prose translation (1408), the Anglo-Norman translation (1271–72), the French translations by Jean de Meun (1284), Jean Priorat (1284–90), and Jean de Vignay (ca. 1315–20), as well as two anonymous French translations (ca. 1280 and 1380) reveals no intermediary source between the Latin Vegetius and our Middle English poem.[32] It is, of course, possible that the source is another French version that has yet to be identified, though Wakelin has observed a close affinity between the writer's composition and its Latin source: "this poem conjures the tone of Latin by using a tortuous grammar, with ablative absolutes and gerundives, and a sesquipedalian phraseology. The vocabulary is even more Latin than the Latin is, with some long words used in English when no like words occur in the original."[33] In sum, our poet freely adapts his source material, sometimes going through it at great speed and with little elaboration. When his source discusses material that surely appeared irrelevant to fifteenth-century practice, such as camels, elephants, and scythed chariots (Veg. 3.23–24), he simply skips over it.[34] In several instances he adds details here and there, or adds entirely new material. Only a few sections receive a great deal of attention and elaboration, most notably Vegetius's treatment of ships and naval warfare (lines 2609–2972), which Dyboski and Arend, rather colorfully, chalk up to the poet having been "a trueborn Englishman."[35]

One of the more important questions to ask about a manual for chivalry and war is how it might have affected the conduct of war. It has been argued by Bernard S. Bachrach, among others, that Vegetius had a practical influence throughout the Middle Ages because narratives describe military leaders following the precepts of his manual.[36] To demonstrate this, he considers only writers who appear to be unaware of Vegetius and narrate war without any rhetorical embellishment, then shows that their descriptions of warfare, strategy, and tactics clearly follow the precepts of Roman manuals. Bachrach claims that military leaders, in line with manuals, attacked only if battle was inevitable, harassed superior forces rather than

---

30 Boffey, "Books and Readers in Calais," p. 70.

31 Reeve, however, facing the nightmare of comparing the freely adapted verse of the text, only went so far in comparing it to Latin Vegetius manuscripts and "ran out of patience" ("Transmission of Vegetius's *Epitoma Rei Militaris*," p. 343).

32 In their edition, Dyboski and Arend only really compared the poem to the Latin Vegetius and Jean de Meun's translation. On the content of these translations see Allmand, *The "De Re Militari" of Vegetius*, pp. 185–87, 152–68 (see also Ed. Galderisi, *Translations médiévales*, 2.256–60). The editions here consulted are: Ed. Lester, *Earliest English Translation*; Ed. Carley, "The Anglo-Norman Vegetius"; Jean de Meun, *Li abregemenz noble honme Vegesce Flave René*, ed. Löfstedt; Jean Priorat, *Li Abrejance de l'Ordre de Chevalerie*, ed. Robert; Jean de Vignay, *Li livres Flave Végèce de la chose de chevalerie*, ed. Löfstedt; *Le livre de l'art de chevalerie et la doctrine de l'enseignement des gens d'armes et à pié et à cheval* (unedited: Sankt-Peterburg, Rossiyskaya natsional'naya biblioteka, MS Fr.f.v.IX 1, fols. 1r–58v; and also in Wolfenbüttel, Herzog August Bibliothek, MS Blankenburg 111, fols. 1r–47r); Ed. Leena Löfstedt et al., *Le livre de l'art de chevalerie de Vegesce.*

33 Wakelin, *Humanism, Reading, and English Literature*, p. 83.

34 Interestingly, the Middle English prose translation, ostensibly more pragmatic than *Of Knyghthode and Bataile*, does not omit but expands on these sections (Ed. Lester, *Earliest English Translation*, pp. 151–54).

35 Ed. Dyboski and Arend, *Knyghthode and Bataile*, p. xxxvi.

36 Bachrach, "Practical Use of Vegetius' *De re militari*."

confront them directly, used surprise, held reserves, positioned their forces so that the enemy faced the sun, held fortifications along lines of supply and communication, used fortifications to deter invasion, and took good care of their horses. Although scholars have identified some "pocket-sized" manuscripts of Vegetius that might be carried on campaign, it is clear that these are merely smaller copies of the text that were meant for use in libraries and not in the field.[37] Until the mid-fourteenth century Vegetius's manual was almost exclusively owned by monks and other religious figures. In the fourteenth and fifteenth centuries the text became more popular with secular owners, especially French military leaders, in a period that witnessed the production of many of the vernacular translations.[38] However, it remains uncertain whether military leaders actually read or used these books at all. The nobility collected large libraries that included such manuals, but in many cases they appear to have used their books only to demonstrate their own military prudence and patronage of mankind's collected wisdom.[39]

It is tempting to imagine military leaders in the Middle Ages making practical use of military manuals. It would allow the often fragmentary, contradictory, and confusing representations of war to be filtered out by comparison. Manuals would answer many of the fundamental questions that recorded sources overlook and reveal how medieval military men approached the problems of war, and so help us to arrive at a clearer understanding of the period. It is evident, however, that those who described war in their narratives did not transparently represent leaders fighting according to Vegetius, but themselves often turned to Vegetius so that they could better write accounts of war. As John Keegan succinctly notes,

> Battles are extremely confusing; and confronted with the need to make sense of something he does not understand, even the cleverest, indeed pre-eminently the cleverest man, realizing his need for a language and metaphor he does not possess, will turn to look at what someone else has already made of a similar set of events as a guide for his own pen.[40]

Since these Roman texts were viewed as authorities, medieval writers employ them to appear well read, lend authority to their writings, and make the military leaders they describe look all the better.[41] However, writers rarely demonstrate their reading of these texts through direct quotation or cited paraphrase, but instead use them in a broader sense to understand warfare. Correlation between Roman precepts and medieval descriptions of war proves only that medieval writers, not military leaders, had read and accepted such ideas. Indeed, chroniclers portray Henry VI as reading Vegetius so that he could appear to be a wise and thoughtful military leader.[42] These manuals offer little in the way of detailed or specialized advice that might be applicable on the medieval battlefield, but rather only "common sense."[43] In the end, even if the text

---

37 For a discussion of these claims, see T. Smith, "National Identity," pp. 29–30.

38 On the ownership of Vegetius manuscripts, see Allmand, *The "De Re Militari" of Vegetius*, pp. 63–80; and Nall, *Reading and War*, pp. 14–36.

39 Allmand notes that about a third of the Latin manuscripts appear unread, and Taylor notes that Charles V owned at least ten French Vegetius manuscripts (Allmand, *The "De Re Militari" of Vegetius*, p. 13; C. Taylor, *Chivalry and the Ideals of Knighthood*, p. 272).

40 Keegan, *Face of Battle*, p. 62.

41 Abels and Morillo, "A Lying Legacy?" p. 11.

42 Wakelin, *Humanism, Reading, and English Literature*, p. 81.

43 See Anglo, "Triumph of Mediocrity."

was not used in a literal sense by English military leaders, it still found importance amongst readers in the period, many of whom were clearly military men, if at least as a codification of the ideals they already held.

But what of our Middle English version of Vegetius? There is more at work in *Of Knyghthode and Bataile* than a simple Middle English translation of Vegetius. The parson's 3,028-line poem was, as the broad political context discussed already implies, deeply connected with its time. As Christopher Allmand observes, the poet's goal seems hardly focused on the accuracy of his translation at all; instead, "his method was more to emphasise certain themes running through Vegetius's work which might be used to build the foundations of a message, social and political as much as military, which would turn his version of *De re militari* into a committed text bearing upon contemporary problems in English society."[44] In this sense, Wakelin writes, though it is "a brilliant verse translation," it should be judged more precisely in terms of a paraphrase: the poet "intersperses Vegetius' dry technical advice with bombastic eulogies of political obedience, and paraphrases much military instruction into mischievous allegories of the possible fate of the king's enemies, the supporters of the Duke of York."[45]

In other areas the text reveals much about the time of transition in which it was composed. In several instances the writer discusses newer technology, most notably gunpowder weapons (such as at lines 1850–51: "gunne and serpentyn that wil not vary, / Fouler, covey, crappaude, and colveryne"), which were used in Europe by 1327 and were becoming more and more important by the mid-fifteenth century.[46] These instances help to illustrate the great variety in gunpowder weapons at this important stage in their development. The poet adds these and other details to bring his text's presentation of warfare up to date to ensure that its readers found it valuable and relevant, rather than just a series of antiquary details.[47]

On the other hand, throughout the text the poet is fairly imprecise with his handling of many of Vegetius's Latin terms.[48] Most notable of these is *miles*, which in classical Latin meant a professional soldier. The word was notoriously slippery throughout the later Middle Ages and, although commonly translated today as "knight," it might refer to a man's military function, equipment, training, experience, or high social status in contrast to other men.[49] Our writer variously translates the word as "chivaler," "knyght," and "werreour," and it is not at all clear what sort of the above-mentioned meanings he might have been aiming for in any given situation, let alone as a whole. Is the poet trying to suggest that knights were meant to be mounted by employing the French term "chivaler," with its equine connotations, or that other combatants were not necessarily of the knightly class by using "werreour"? Indeed, the Latin term *bellator* (meaning "warrior" or "fighter") is variously translated as "chivaler," "bellatour," and "werreour," with no suggestion that these were distinct from "werreour" as translated from *miles*. The difficulty in terminology is clearly expressed in the following passage (lines 1209–15):

44 Allmand, "English Translations," p. 4.

45 Wakelin, "Occasion, Author, and Readers," p. 260.

46 See Rogers, "Gunpowder Artillery in Europe."

47 Allmand, "Fifteenth-Century English Versions," p. 43. Other medieval writers who made use of Vegetius, such as Giles of Rome, felt it necessary to update their texts as well, on which see Contamine, "Les Traités de Guerre," pp. 354–55.

48 For tables comparing use of vocabulary in different translations, see Allmand, *The "De Re Militari" of Vegetius*, pp. 350–53; Ed. Dyboski and Arend, *Knyghthode and Bataile*, pp. lviii–lxxiii. See also Allmand's discussion in "Fifteenth-Century English Versions," pp. 35–38.

49 See, for example, Prestwich, *Armies and Warfare in the Middle Ages*, pp. 12–18.

The chivaler, be he legionary, — *knight, whether he [is a] legionnaire*
 As seide it is beforn, on hors or foote,
Or aydaunt, that is auxiliary,
 On hors or foot — if that thei talk or mote — *discuss*
Of werre, and reyse roore, up by the roote — *war, and raise riot*
 Hit shal be pulde with myghti exercise — *pulled*
 Of werreourys, governed in this wise.

Here we see a "chivaler," which may or may not also be a "legionary," either on horseback or on foot, in opposition to an "aydaunt" (auxiliary), either on horseback or on foot, and then "werreourys," seemingly employed in a more general sense for all combatants. In most cases it is not clear whether he employs different translations to signify different types of men, or merely to fit the meter or rhyme. Allmand succinctly notes that "There appears to have been little attempt at consistency here. Was there ever intended to be such? Are we too 'modern' in expecting it? Probably so."[50]

In other cases, the direct senses of the Latin words are more clearly translated for their inherent functions. The Latin *pedites* (meaning those on foot) is translated straightforwardly as "footmen" or "men on foote." *Equites* (meaning those on horseback) is translated as "hors" (similar to the Napoleonic "horse" for cavalry), "horsemen," and "ryderys." These two ideas, in their varied translations, were often set in direct contrast with each other, and so shows that the poet viewed them as words meant to convey modes of fighting, rather than anything related to quality or equipment.[51] But such cases of clear meaning in the poem are rare when it comes to combatants.

The often fluid understanding of these terms for military men further reflects the ever changing makeup of armies in this period, when the use of the heavily armored combatant, often of the knightly class, had diminishing importance in the face of increasing reliance on missile weapons (gunpowder or otherwise). Through the fourteenth and fifteenth centuries the proportion of men-at-arms to archers steadily declined in English forces.[52] Men-at-arms were generally from landed families, but were not necessarily of the knightly class. Indeed, fewer and fewer knights opted to fulfill their service when called. The poet, unlike his source, focuses on the importance of noble birth (for example at lines 271–77), although he does not use "knyght" to consistently signal higher social status. This reflects the late-medieval debates on whether nobility, and thus meritorious status, was gained from birth or earned through conduct.[53] Archers were generally from the lower stratas of society, although they were on occasion members of gentry families, despite the poem's suggestion that they were drawn from the same pool of men as knights (lines 432–34).[54] Armies also included more specialists, especially to man gunpowder weapons, which saw increasing use throughout the fifteenth century.

The poem's attention to other areas reveals much else about the poet's concerns. The focus on discipline, logistics, and the importance of paying one's military men (such as at lines 278–80, 397–99, 483, 603–06, and 796–98), is part of a larger dialogue in English writings in this period after the Hundred Years

50 Allmand, "Fifteenth-Century English Versions," p. 37.

51 See lines 552, 601–02, 644, 654, 656–58, 717–18, 730–32, 748–49, 1367–68, 1398–99, 1558, 1578, 1599–1600, 1760, 1839, 1902–03, 1956, 1958, 2077–78, 2082, 2087, and 2203–04.

52 Bell et al., *Soldier in Later Medieval England*, pp. 59–72, 260–74.

53 Vale, *War and Chivalry*, pp. 14–32.

54 Baker, "Socio-Economic Origins of English Archers."

War (1337–1453) that attempts to understand why the English were defeated.[55] This focus might have been influenced by the poet's experiences in Calais, where the members of the garrison were particularly fickle when it came to their pay. Other details on kingship and leadership found throughout suggest an anxiety over the instability that England was suffering.[56] The omission of some details, notably all mention of retreat, along with a considerable shift in tone and style, with battles "written in high imaginative excitement," all reveal a far more "chivalrous" interpretation of war.[57] Although the French versions of Vegetius typically include the word "chivalry" in the title and their discussions, they rarely discuss the ideals of knighthood and knights, and instead mostly present sober translations.

MANUSCRIPT HISTORY AND PROVENANCE

Of Knyghthode and Bataile is indexed as item 3185 in ed. Boffey and Edwards, *New Index of Middle English Verse,* and it survives in four known copies:

- MS: Cambridge, Pembroke College, MS 243, fols. 1r–55v. [Base-text for this edition.]
- A: Oxford, Bodleian Library, MS Ashmole 45 (Part 2), fols. 1r–7v, 18r–23v, 41r–43v, 46r–53v.
- C: London, British Library, MS Cotton Titus A.xxiii, fols. 2r–53v.
- R: London, College of Arms, MS R.25, fols. 24r–62v.[58] [This text not in NIMEV.]

The text is transmitted fairly accurately in MS, A, and C, although there are some changes in A and C, and many more in R. None of the four copies appear to have been originally bound with any other material, although A and R have since been rebound with other texts.[59]

MS is an octavo volume with vellum pages of 241 x 166 mm, with the writing occupying 164 x 95 mm of space, that has post-medieval binding. It is written in Secretary with some Anglicana forms in a hand of the third quarter of the fifteenth century. The first letters of some words, especially at line beginnings, are rubricated, and certain other words are written entirely in red ink. The first letter of each of the proems and the books are many lines in height and heavily decorated, typically with red and blue ink. The volume is probably not an autograph copy, although it is the earliest of the four texts, as it includes praise of Henry VI on several occasions and does not have the many changes found in the later manuscripts. In its early years it may have been owned by the family of William and Ralph Hastings.[60] Its text is described in greater detail below.

C, written in a hand contemporary to that of MS, was first owned by one Edward Hatcliff, whose name is inscribed on fols. 56v and 57r.[61] It must have been written after July 10, 1460, when John Beaumont was

---

55 See Nall, "Perceptions of Financial Mismanagement" and *Reading and War*, pp. 48–74.

56 Allmand, "English Translations," pp. 4–5.

57 Ed. Dyboski and Arend, *Knyghthode and Bataile,* p. xxxii. See also Whetham, *Just Wars and Moral Victories,* pp. 159–60.

58 R includes an unfoliated leaf after fol. 38.

59 For codicological descriptions of MS, A, and C, see Nall, "Production and Reception of Military Texts," pp. 285–87, and for R, see Davies and Edwards, "New Manuscript of *Knyghthode and Bataile*." Dyboski and Arend largely skip over the manuscripts themselves, and even go so far as to print out the published catalogue entries for MS, A, and C instead of reassessing them anew (*Knyghthode and Bataile,* pp. xi–xiv).

60 Nall, *Reading and War,* p. 34.

61 On the identity of Hatcliff, see Boffey, "Books and Readers in Calais," p. 69n12.

killed and Henry VI was captured at the battle of Northampton, because it suppresses all mention of Henry VI in a positive light, notably by omitting both the general proem (lines 1–88) and the proem to Part III (lines 978–1026), while also replacing Henry VI's name with Edward IV's at lines 121 and 2880, and significantly changing the meaning of lines 2982–85 to fit a different ideal of war.[62] It omits the Latin passages at the beginning of Parts II and IV (before lines 621 and 2224), possibly because the other two Latin passages were then absent when the general proem and proem to Part III, which they were within, had been removed. C also includes glosses above some peculiar or imprecise terminology, mostly in English but also occasionally in Latin, and an alphabetical index (fols. 54r–56v), rare for Middle English texts.[63] Such glosses of military vocabulary were often added by English humanists to classical writings, in both their original and translated forms.[64] As Wakelin has demonstrated, these changes to the text were in the exemplar that C was copied from.[65]

A was itself copied from C not long after it was written.[66] The two texts are very close throughout, with essentially the same glosses and index (fols. 54r–56v), and are even in very similar hands, although a full comparison is impossible due to the many leaves that have been lost from A over the centuries. (This loss had begun by the late fifteenth century at the latest, as fols. 15–16 were evidently missing by then, when R was written.) The changes related to Henry VI and Edward IV are all present (lines 121, 2880, and 2982–85, and the foliation of the MS indicates that lines 1–88 and 978–1026 were omitted following C), and the Latin passages are likewise absent.

R was possibly owned early on by somebody named "John," as the name is written on fols. 30v and 34v. It was written in the late fifteenth century and follows C and A in replacing praise of Henry VI with Edward IV, omitting the two sections of the text that were in favor of the Lancastrian king, and reordering phrases in many other instances. A word-by-word comparison of R with MS, A, and C indicates that R was copied from A.[67] Setting aside spelling, C has six differences from our base text that are not followed by A nor R:

Line 140	has *in* in place of *and*
Line 1321	adds *joo jo* before *journal* (uncancelled dittography)
Line 1341	has *of* in place of second *and*
Line 1428	adds *is* before *is* (uncancelled dittography)
Line 2674	omits *tan*
Line 2925	omits *to make*

62 For these key variants, see the Textual Notes.

63 Dyboski and Arend erroneously claim that some of these glosses are in MS, and state that "it is difficult to see what other aim they could have had than the mere gratification of personal vanity," when in reality it is clear that the scribe of C was trying to clarify what exactly he thought these words meant (*Knyghthode and Bataile*, p. xxvi).

64 Wakelin, *Humanism, Reading, and English Literature*, p. 83.

65 Wakelin, "Scholarly Scribes," pp. 29–30.

66 On A's relation to MS and C, see Wakelin, "Scholarly Scribes," pp. 32–36. Dyboski and Arend, however, believe that A and C have a common source different from MS (*Knyghthode and Bataile*, p. xiv).

67 Although Davies and Edwards ("New Manuscript of *Knyghthode and Bataile*") rightly note that R follows the Yorkist changes of C and A, they make no attempts to identify which of the two is closest to R, no doubt because of the very close similarity of C and A. It is therefore necessary to provide the information below in detail.

In each of these cases R follows not C, but A, which has silently corrected these changes. In three other instances R follows the visible corrections that the scribe of A made to his text after copying it from C:

Line 2322	C has *sadde* in place of *saddest*, while A has *sadde* with *st* added in afterwards to make *saddest*
Line 2422	C has *hevy* in place of *every*, while A has *hevy* with the *h* scraped and an *er* abbreviation added afterwards to make *every*
Line 2988	C omits *that*, while A has *this* in place of *that*

In only one case does R follow C against A (at line 1194 they omit *tech*), but this is certainly because the scribe accidentally passed over A's reinsertion of the missing word in superscript. Indeed, the scribe of R was not as careful as those of C and A, and has many further changes to the wording and word order of his text throughout. Some of the more peculiar alterations are due to his misreading of his source manuscript, such as *bataile* for *vitaile* twice (lines 1066 and 1481), *oragge* for *cragge* (line 2250), *soon* for *foon* (line 1784), *with* for *wight* (line 1832), and an unabbreviated *the* for *ye* ten times because of his confusion between his source's formation of *þ* (for *th*) and *y* (lines 401, 1161, 1162, 1430, 1572, 1948, 2511, 2822, 2844, and 2976). The most egregious instance of his mangling of the text is when he wrote the first two words of line 2491 but then, due to eye skip, filled out the line with line 2492 from the second word onwards.[68] He noticed his error almost immediately and attempted to fix it by cancelling the first two words of line 2491, replacing them with the first word of line 2492 in superscript, and then adding in line 2491 in full after line 2494, where it creates a peculiarly ordered rhyme scheme (*aabbbcc*). At other times he skips over individual lines (465 and 1114) and once an entire stanza (607–13), seemingly by accident.[69] Besides these lines and those omitted from A and its source, C, he omits sixteen consecutive stanzas (lines 866–977), which are equivalent to two leaves that were originally in, but are now missing from A (fols. 15–16), his source manuscript. R has many other changes to its text throughout, although the substance is much the same.

MS employs a consistently structured layout. For many of the stanzas it includes short Latin summaries, many of which appear to be taken from the Latin Vegetius. Its general proem (lines 1–88) is in eight-line stanzas of *ababbcbc*, with the final rhyme of each stanza rhyming with the first of the following stanza. The rest of the text is in seven-line stanzas of *ababbcc* (rhyme royal), although the proems to Parts II, III, and IV (lines 621–41, 978–1026, and 2224–44) also have the final rhymes of each of their stanzas rhyming with the first of those that follow, just like the general proem. Aside from the proems, the text is arranged with eight stanzas to each leaf, with four to a side, lettered A–D on the rectos and E–H on the versos. The manuscript's main text, omitting the general proem, was foliated when it was originally written, although early in the manuscript's history it was rebound so that several of its bifolia are now ordered incorrectly.[70]

---

[68] Davies and Edwards ("New Manuscript of *Knyghthode and Bataile*," p. 139), however, suggest that this error was due to the scribe changing his mind while writing the line.

[69] Davies and Edwards ("New Manuscript of *Knyghthode and Bataile*," pp. 140–41n10) propose that the omission of lines 607–13 was intentional and was meant to allow Part II to commence at the top of fol. 31v.

[70] MS has since been refoliated in its current disordered state, starting not at the main text but at the general proems, but if it were to be rearranged, in its current numbering we would have fols. 43 (incorrect) as 42 (correct), 44 as 43, 42 as 44, 47 as 45, 45 as 46, and 46 as 47. We owe thanks to James Freeman of Cambridge University Library for

At the conclusion of his poem, the poet asks copiers to respect the text: "Thi writer eek, pray him to taken hede / Of thi cadence and kepe ortographie, / That neither he take of ner multiplye" (lines 3026–28), which clearly influenced the C and A scribes, as they closely follow the text (often preserving peculiar spellings) and its layout.[71] C was originally foliated in exactly the same manner as MS, but skips folio 17 in its foliation because the manuscript from which the scribe was copying was missing this leaf (lines 978–1026, at the modern foliation of fol. 19 in MS, originally fol. 17), certainly because its text here was strongly in favor of Henry VI.[72] A, copied from C, has the same layout and retains its original foliation, although many of its leaves are now missing (fols. 8–17, 24–40, and 44–45, the equivalent of lines 481–1026, 1363–2300, and 2469–2580). In many instances a different hand has corrected its text against another that is very close to MS. The text's layout, with folio numbering and stanza lettering, enabled the reader to quickly identify and reference the section they were examining, and allowed for precise references to be included in the indices at the end of C and A.[73] The layout, along with the glossing and attention to detail, suggests that the text was meant to be read and extensively used. This consistency in layout and a conscious attempt to preserve the text reveals the humanist influences on the writer and these scribes.[74] The R scribe, in contrast, largely ignored the original layout of the text. He added an additional stanza per page, making ten per leaf, and omitted the rest of the apparatus in his source, including the glosses, stanza lettering, and index.

EDITORIAL PRACTICE

MS is used as the base text for this edition because it offers the earliest and fullest version of the text. Major variants in A, C, and R are included in the Textual Notes, but minor differences, such as spelling, are omitted. The text has been corrected in the few instances where there are clear errors, with each case detailed in the Textual Notes. We note textual deletions with struck-through text in the Textual Notes, but make no distinction between different types of deletion. In instances where the hands are rather ambiguous we have marked our uncertainty with question marks (?). As mentioned above, there are marginalia and glosses in the manuscripts, but because these are typically repetitions of the text's meaning they have been silently omitted. In a few instances in which these help to understand the meaning of the text, especially for its peculiar terminology, they are cited and discussed in our notes. In our marginal glosses we have tried to retain the peculiarities of the poet's terminology, regardless of the original Latin words that he translates from, and we provide different glosses for each term whenever applicable. This has helped us to avoid implying any sort of additional professionalism that did not yet exist in the period, as would have been suggested by using the simplified terms of "soldier," "infantry," or "cavalry."

The earlier edition of the text has many issues that merit this new edition, no small number of which result from its troubled production history.[75] Roman Dyboski, the initial editor, began work on his edition

sharing his thorough analysis of the manuscript with us. Foliation utilized in this volume will follow this corrected order rather than the current foliation.

71 Among other factors, scribal attempts to retain the orthography of source manuscripts, and their varying muddled results, makes it difficult to determine their origins through linguistic means.

72 C has also been refoliated to start at fol. 2. For evidence that fol. 17 and its text were missing from the manuscript that C was copied from, see Wakelin, "Scholarly Scribes," p. 30.

73 Interestingly, a very similar layout is employed in *On Husbondrie*, the text that others have argued was also composed by our poet. See Wakelin, "Scholarly Scribes," p. 28.

74 Wakelin, "Scholarly Scribes."

75 Ed. Dyboski and Arend, *Knyghthode and Bataile*, p. vii. See, further, Coleman, "Roman Dyboski."

in 1906. He soon transcribed the three texts that he knew of, and made progress in other areas, but did not have the luxury of being able to easily revisit (let alone photograph!) his manuscripts to double-check aspects of them that he may have mis-transcribed or been otherwise unsure about. His pace was slowed by national events, notably his participation in the First World War and his subsequent seven years in Russian incarceration. Some time after he regained his freedom he took on his colleague, Zygfryd Marjan Arend, as a co-editor, and together they finished the volume in 1935. The sporadic work on their edition encouraged much inconsistency and unevenness in the treatment of the text and the varied subjects discussed in the volume. Besides their focus on the text's language, their major concern in examining the text was to see how it related to the Latin Vegetius.[76] They did not properly treat it as a text with its own cultural, intellectual, literary, and historical contexts. The aim of the present edition of *Knyghthode and Bataile* is to provide both a more accurate text and a more contextualizing apparatus to aid the reader.

We have also endeavored to follow the base manuscript more closely whenever possible. Here, for instance, is our text, lines 1657–63 of the poem, in a transcription from the base manuscript (fol. 31r), followed by the edited text in both the Early English Texts Society edition of Dyboski and Arend and the present edition. The transcription:

> At brigge or hard passage or hillis browe
> Is good to falle vppon · Or if ther be
> mire or mareys · or woode or grovis rowe
> or aggravaunt other difficultee,
> To falle uppon is thenne utilitee
> The hors to sech · Vnarmed ar aslepe /
> To falle vppon is good to take kepe ·

Dyboski and Arend:

> At brigge or hard passage, or hillis browe,
> Is good to falle vppon; or if ther be
> Mire or mareys or woode or grovis rowe
> Or aggravaunt other difficultee,
> To falle vppon is thenne utilitee;
> The hors to sech vnarmed <u>or</u> aslepe
> To falle vppon is good to take kepe.

[76] See, for instance, their rather critical introductory remarks: "[the poet] frequently selects his material from the text before him in quite an arbitrary manner, so as to make it suit the exigencies of the metre, and either abandons the rejected details altogether, or introduces them in another connexion later on; thus the pace in which he follows the footsteps of Vegetius, often shuffling even to tediousness, sometimes changes to an irregular ramble backwards and forwards over a whole series of chapters or paragraphs; this is coupled with very common misunderstandings of the original, most of them due to the author's practical inexperience in military matters, from which he often seeks refuge in clumsy generalities and commonplace tags" (Ed. Dyboski and Arend, *Knyghthode and Bataile*, p. xxxiii).

Present edition:

At brigge or hard passage or hillis browe
Is good to falle uppon. Or if ther be
Mire or mareys, or woode or grovis rowe
Or aggravaunt other difficultee,
To falle uppon is thenne utilitee
The hors to sech. Unarmed ar aslepe;
To falle uppon is good to take kepe.

The reader will note small punctuation differences in the initial lines, where the present edition attempts to follow the MS indicators more closely. The most significant differences, however, occur in line 1662, where the misreading by Dyboski and Arend of MS *ar* as *or* (underscored above) reverberates across the MS punctuation: in search of a verb, the editors are forced to ignore both the mid-line punctus and end-line virgule in that line. Their result, for which they provide little rationale, garbles the meaning of the line whether compared to the poet's apparent intent or his source in Vegetius.

As a more complicated and fascinating example, here are lines 2980–86 of the poem, first in transcription (fol. 55r):

Hail porte saluz · with thi pleasaunt accesse
Alhail Caleis · Ther wolde I faynest londe
That may not Ioo · Whi so · For thei distresse
alle / or to deye / or with her wrong to stonde
That wil I not to wynne al Engelonde
What myght availe · A litil heer to dwelle /
And world withouten ende abide in helle ·

Dyboski and Arend:

Hail, porte saluz! with thi pleasaunt accesse,
Alhail Caleis! ther wolde I faynest londe;
That may not I [—] oo, whi so? for thei distresse
Alle, or to deye or with her wrong to stonde.
That wil I not, to wynnne al Engelonde!
What myght availe, a litil heer to dwelle,
And world withouten ende abide in helle.

Present edition:

Hail, porte saluz! With thi pleasaunt accesse,
Al hail Caleis! Ther wolde I faynest londe.
That may not Joon! Whi so? For thei distresse
Alle, or to deye, or with her wrong to stonde.
That wil I not to wynne al Engelonde!

> What myght availe? A litil heer to dwelle,
> And world withouten ende abide in Helle!

Again, there are minor changes of punctuation scattered through the passage, but the reading of line 2982 (underscored) calls out for particular attention. The base manuscript records a clear *Ioo* or *Joo* (medieval *I* and *J* being interchangeable), but no reading of *Ioo/Joo* seems to withstand scrutiny within the context of the line. To remedy the situation, Dyboski and Arend suggest that these are two words. Though the scribes in these manuscripts do have a general, if not perfect, respect for the spaces between words, adding such a space is not a wholly unacceptable emendation on its own, just as it is in adding a space between the words *Al* and *hail* in the previous line. However, Dyboski and Arend further add an intervening mark of punctuation via an em-dash that, while bracketed in their text, is inexplicably labeled as being "in MS" within their footnotes.[77] As the transcription above shows, there is no punctuation between the three closely aligned letters. Nevertheless, the editorial invention of Dyboski and Arend has subsequently led critics down a rabbit hole of creative readings. Wakelin, for instance, appears to have taken their em-dash as a purposeful omission, such that he suggests a far different punctuation — *That may not I . . . oo, whi so?* — and thereby wonders if this odd line "deliberately omits the poet's name."[78]

Examination of the other three manuscripts only complicates the situation, as this stanza has been heavily altered in all of them, as seen in a transcription of C (which is closely followed by both A and R):

> Hayle poort saluz with thy plesaunt accesse
> al hail Caleys · There wold I fayne o londe
> That maynot Ioo · whiso · for they distresse
> alle or to deye / or with here werke to stonde
> That dar to right go wynne alle Engelonde
> what myght availe a lite in errour dwelle /
> and world withouten ende abide in helle

These changes are no doubt related to the previously discussed shift in the political history surrounding the text. The original, Lancastrian text, was written at a time when Calais was in the possession of the rival Yorkists. Thus the poet in this stanza complains that he is exiled from Calais, which is in the hands of those (i.e., the Yorkists) who would either kill him or make him stand with their wrong (i.e., their rebellion). He insists he would not help this enemy even if it personally earned him possession of all England, because the result would only be a temporary victory: life in this world is finite, and such a betrayal would earn him an eternity in Hell. Such an anti-Yorkist stance was no longer suitable with the Yorkists in control of both England and Calais. The Yorkist revisor of the poem thus had to do something with the stanza, and his revisions may have been led by the original MS *Ioo/Joo*. Facing the same uncertainty as modern editors in determining its meaning, he seems to have read *Joo* as meaning 'Jew'. This reading replaces the Yorkist rebels with Jews, who by law could not be in Calais.[79] Further, the text would then suggest that the Jews would be

77 This is not unusual within the Early English Texts Society edition. As a work of co-authorship written over some years (see pp. 14–15, above), its notes and text are frequently at odds with one another.

78 Wakelin, "Occasion, Author, and Readers," p. 265.

79 Edward I had expelled the Jews from England in 1290, and they were not allowed to return to England until 1657. In the intervening period, the recorded Jewish presence in England (and its holdings, like Calais) is largely con-

unwelcome regardless, as they would either kill the citizenry or forcibly convert them to the *werke* of their faith, which would attempt to seize all England. These Jews will live here in error (i.e., sin) and then earn an eternity in Hell.

If an anti-Semitic reading is possible within the subsequent texts, could it be original to the poem? Abrupt though this turn to anti-Semitism would be, it would not be wholly impossible from the self-described parson who throughout *Of Knyghthode and Bataile* is quick to condemn his enemies to torment and had, in the previous stanza, praised the Virgin Mary. Still, it seems highly improbable given the consistent anti-Yorkist positions in the text and how well a simple political reading fits within the context of the poem here — the next stanza looks forward to a time when the king would regain governance of Calais — and elsewhere. What, then, is the editor to do with MS *Ioo/Joo*? The suggestion here is that the confusion is the result of a scribal omission of a macron (the scribe makes a similar mistake in writing *oo* for *oon* in line 2403). MS *Joo* would then be read as *Joon*, an odd spelling for the name John. While it would be tempting to read this name as that of the poet himself, John frequently stands as a generic reference to a priest or man in the Middle Ages (as it remains today in "John Doe"). Rather than signing the poem or intending to engage in anti-Semitism, the poet would simply be noting that a fellow like him cannot go to Calais . . . which was precisely the case.

In sum, separation between this edition and its forebear are numerous. Beyond differences in the reading of the manuscript and the related editorial punctuation of its lines, this edition takes a more compassionate look at the poet and his work. The previous editors, for example, often dismiss the poet as mistaken in his understanding of Vegetius, both "untechnical" and "clumsy" (see, e.g., our Explanatory Notes to lines 1734–40, 1748–49, and 2968), when they themselves appear to be in error. There is little doubt that the compound effect of these judgments has negatively impacted our ability to understand and appreciate this text. We hope that this present edition is therefore more sensitive to *Of Knyghthode and Bataile*'s craft, and more respectful of its poet's contemporary reading of Vegetius.

fined to occasional foreign visitors and a small number of Jews who converted to Christianity and were given residence in the Domus Conversorum in London.

Proem

Proemium

fol. 1r	*Salve, festa dies I Martis, Mavortis! Avete Kalende!*	
	Qua Deus ad celum sublevat ire David!	*(See next three lines)*
	Hail Haly Day devout! Al hail kalende	*Holy Day; kalends (i.e., first day)*
	Of Marche, wheryn David the Confessour	
	Commaunded is his Kyngis court ascende![1]	
	Emanuel, Jhesus the Conquerour,	
	This same day as a tryumphatour,	*supreme victor*
	Sette in a chare and throne of majestee,	*chair*
	To London is comyn. O Saviour,	
	Welcome a thousand fold to Thi citee!	
	And she, Thi Modir (blessed mot she be!)	*Mother (i.e., Mary); blessed may*
	That cometh eke, and angelys anende,	*also, and angels constantly*
	Wel-wynged and wel-horsed, hidir fle,	*flying here*
	Thousendys on this goode approche attende,	
	And ordir aftir ordir thei commende,	*order; they salute*
	As seraphin, as cherubyn, as throne,	*seraphim; cherubim; thrones*
	As domynaunce, and princys hidir sende.	*dominions; principalities*
	And, at o woord, right welcom everychone!	*one*
fol. 1v	But Kyng Herry the Sexte, as Goddes Sone	*Henry VI, like*
	Or th'emperour or Kyng Emanuel,	
	To London, welcomer be noo persone!	*more welcome is no one [but you]*
	O souverayn Lord, welcom! Now wel, now wel!	
	Te Deum to be songen, wil do wel,	*'Thee, O God [we praise]'*
	And *Benedicta Sancta Trinitas*!	*'Blessed [be] the Holy Trinity'*
	Now prosperaunce and peax perpetuel	*prosperity and perpetual peace*
	Shal growe. And why? For here is *Unitas*.	*'Unity'*
	Therof to the Unitee: *Deo gratias*	*'Thanks be to God'*
	In Trinitee! The clergys and knyghthode	*clergy; knights*
	And comynaltee better accorded nas	*commons better accorded were*

1 *Was commanded [to] ascend [into] his King's (i.e., God's) court*

	Never then now. Now nys ther noon abode,	*Now there is no one facing off*
	But out on hem that fordoon Goddes forbode,[2]	
	Perjurous ar, rebellous, and atteynte,	*criminal*
	So forfaytinge her lyif and lyvelode,	*forfeiting their lives and livelihood*
	Although ypocrisie her faytys peynte.	*hypocrisy disguises their fates*
	Now, person of Caleys, pray every seynte	*parson of Calais, pray [to]*
	In hevenys and in erth of help: th'availe	*on earth for help: the benefit*
	It is. That in this werk nothing ne feynte,	*nothing be held back*
	But that beforn good wynde it go ful sayle;	*before; sail*
	And that not oonly prayer, but travaile	*labor*
	Heron be sette. Enserche and faste inquere[3]	
	Thi litil book *Of Knyghthode and Bataile*	*Your; Battle*
	What chivaler is best, on it bewere.	*knight; be advised*

[Conversation between the Poet and Beaumont]

fol. 2r	Whil *Te Deum laudamus* up goth there	*'Thee, O God we praise' is sung*
	At Paulis, up to Westmynster go thee;	*Paul's (i.e., St. Paul's Cathedral)*
	The kyng comyng, *honor, virtus* the quene,	*[with] 'virtue' the queen*
	So glad goth up that blisse it is to see.	*Such happiness arises*
	Thi bille unto the kyng is red, and he	*Your petition; read*
	Content withal, and wil it not foryete.	
	What seith my lord Beaumont? "Preste, unto me	*Priest*
	Welcom." Here is t'assay, entre to gete!	*Now is the test, [with] entry to gain*
	"Of knyghthode and bataile, my lord, as trete	*as is treated in*
	The bookys olde, a werk is made now late,	*a poem was made of late*
	And if it please you, it may be gete."	*fetched*
	"What werk is it?" "Vegetius translate	*translated*
	Into balade." "O preste, I pray thee, late	*Into poetry; let*
	Me se that werk." "Therto wil I you wise.	*I will reveal it to you*
	Lo, here it is!" Anoon he gan therate	*At once he began of it*
	To rede, thus: "Sumtyme it was the gise. . ."	*read; fashion*
	And red therof a part. "For my servyse	*read*
	Heer wil I rede," he seith, "as o psaultier."	*one psalter*
	"It pleaseth you right wel. Wil your advyse	*As it pleases you*
	Suppose that the kyng heryn pleasier	*Whether; pleasure of this*
	May have?" "I wil considir the matier.	
	I fynde it is right good and pertynente	

2 *Except out among those who frustrate God's prohibition*

3 *Be set upon this task. Search and surely inquire within*

Unto the kyng. His celsitude is hier; — *highness is here*
I halde it wel doon, hym therwith presente." — *consider it well done*

[PRESENTATION OF THE POEM TO THE KING]

fol. 2v "Almyghti Maker of the firmament,
O mervailous in every creature,
So singuler in this most excellent — *So uniquely present within*
Persone, our souverayn lord! Of what stature — *(i.e., Henry VI)*
Is he, what visagynge, how fair-feture, — *visage; fair-featured*
How myghti mad, and how strong in travaile! — *mightily made; works*
In oonly God and hym it is t'assure — *to affirm*
As in a might, that noo wight dar assaile. — *mighty one, that no person dare assail*

"Lo, souverayn lord, *Of Knyghthode and Bataile*,
This litil werk, your humble oratour,
Ye, therwithal your chivalers, t'availe,
Inwith your hert to Crist the Conquerour,
Offreth. For ye theryn doeth him th'onour.[4]
His true thought, accepte it, he besecheth. — *he (i.e., the poet) begs*
Accepte it is to this tryumphatour
That myghti werre exemplifying techeth." — *war teaches by examples*

He redeth, and fro poynt to poynt he secheth, — *reads [it]; searches [it]*
How hath be doon, and what is now to done;[5]
His providence on aftirward he strecheth,
By see and lond. He wil provide sone — *sea and land; [forces] soon*
To chace his adversaryes everychone;
Thei hem by land, thei hem by see asseyle.[6]
The kyng his oratoure, God graunt his bone, — *orator; prayer*
Ay to prevaile in knyghthode and bataile. — *Always*

Amen.

[4] Lines 73–77: *Sovereign lord, your humble petitioner offers to you this little work, 'Of Knighthood and Battle', to profit you, along with all your knights, that you might offer your heart to Christ the Conqueror. For you thereby do him (the poet or Christ) honor*

[5] *What had been done in the past [regarding the military], and what is to be done now*

[6] *They (i.e., the king's forces) to assail them (the enemy) by land and sea*

Part I: Recruitment and Training

fol. 3r Of Knyghthode and Bataile

[Military Manuals (Veg. 1.preface)]

Sumtyme it was the gise among the wise	*Once; manner*
To rede and write goode and myghti thingis,	
And have therof the dede in exercise.	*deed exercised [in the mind]*
Pleasaunce heryn hadde emperour and kingis.	*Pleasure in this*
O Jesse flour, whos swete odour our kinge is,	*Jesse's flower (i.e., Christ)*
Do me to write of knyghthode and bataile	
To Thin honour and chivalers t'availe.	*Your; to profit knights*

[Invocation of Christian Militarism]

Mankyndys lyfe is mylitatioun,	*strife*
And she, thi wife, is named *Militaunce*	
Ecclesia. Jhesu, Salvatioun,	*(i.e., the Church Militant)*
My poore witte in Thi richesse avaunce,	*through Your riches advance*
Cast out therof the cloude of ignoraunce,	
Sette up theryn Thiself, the verrey light,	*true*
Therby to se Thi militaunce aright.	*militarism*

O Lady myn, Maria, lode-sterre,	*lodestar*
Condite it out of myst and nyght, that dark is,	*Lead*
To write of al by see and lond the werre.	*all the war by sea and land*
Help, angelys, of knyghthode ye ierarkys	*hierarchies*
In heven and here. O puissaunt patriarkys,	*powerful patriarchs*
Your valiaunce and werre in see and londe	*wars at sea and on land*
Remembering, to this werk putte your honde.	*hand*

Apostolys, ye, with th'almyghti swoorde	
Of Goddis woord, that were conquerourys	*God's*
Of al the world, and with the same woorde	

Ye martirys that putte off sharpe shourys,[1]
Ye virgynys pleasaunt and confessourys *virgins*
That with the same sworde have had victory,
Help heer to make of werre a good memory. *account*

fol. 3v And every werreour wil I beseche, *warrior; beg*
Impropurly where of myn ignoraunce *[If] incorrectly out of*
Of werre I write, as putte in propre speche *[that you] put it more properly*
And mende me, prayinge herof pleasaunce *correct*
To God be first, by Harry, kyng of Fraunce *Henry VI*
And Englond, and thenne ereither londe, *then in either*
Peasibilly, that God putte in his honde. *Peacefully*

Thus seide an humble invocatioun
To Criste, His Modir, and His sayntis alle:
With confidence of illustratioun, *inspiration*
Criste me to spede, and prayer me to walle, *to assist me; to protect me*
Myn inwit on this werk wil I let falle, *My mind*
And sey what is kynyghthode, and in bataile,
By lond and see, what feat may best prevaile.

[Definition of Knighthood]

Knyghthode an ordir is, the premynent; *preeminent*
Obeysaunt in God, and rather deye *Obedient to; die*
Then disobeye; and as magnificent *exulted*
As can be thought; exiled al envye;
As confident the right to magnifie
As wil the lawe of Goddis maundement, *God's commandment*
And as perseveraunt and patient. *persevering*

[Celestial Example of Knighthood]

The premynent is first th'Almyghti Lord,
Emanuel, that every lord is undir
And good lyver; but bataile and discord *righteous person*
With him hath Sathanas: thei are asondir *Satan; they are as separated*
As day and nyght, and as fier wasteth tundir, *fire destroys tinder*
So Sathanas his flok; and Cristis ooste *Satan [destroys] his flock; Christ's army*
In gemmy gold goth ardent, every cooste. *bejeweled; goes gleaming; coast*

[1] *You martyrs who put off bitter attacks*

fol. 4r Th'Emanuel, this Lord of Sabaoth, — *Heavenly hosts*
Hath ostis angelik that multitude, — *Has hosts of angels*
That noon of hem, nor persone erthly, wote — *not one of them; earthly, knows*
Their numbir or vertue or pulcritude.
Our chivalers of hem similitude
Take as thei may, but truely that ful fer is,
As gemmys are ymagyned to sterrys.[2]

Folk angelik, knyghthode archangelike, — *knights are like archangels*
And the terrible tourmys pryncipaunt, — *ruling troops [are] principalities*
The potestates myght, ho may be like, — *powers' strength, they*
The vigoroux vertu so valyaunt, — *vigorous*
The regalye of th'ordir domynaunt, — *dominions*
The thronys celsitude of cherubyn. — *thrones' highness of cherubim*
Who hath the light or flamme of seraphyn? — *seraphim*

Yit true it is, man shal ben angelike;
Forthi their hosteyinge the Lord hath shewed — *Therefore their war-waging*
Ofte unto man: the crafte therof to pike, — *the art of it thereby to learn*
In knyghthode aftir hem man to be thewed. — *instructed*
By Lucyfer falling, rebate and fewed — *reduced and made small*
Her numbir was, and it is Goddis wille, — *Their; God's will*
That myghti men her numbir shal fulfille. — *replenish*

[CONTENTS OF THE BOOK (VEG. SYNOPSIS)]

Of myghty men first is th'electioun — *selection*
To make, and hem to lerne and exercise;
An ooste of hem for his perfectioun, — *army*
Be numbred thenne; and aftir se the gise — *see the manner*
Of strong bataile, fighting in dyvers wise;
In craft to bilde, and art to make engyne — *build; engines [of war]*
For see and lond, this tretys I wil fyne. — *treatise; complete*

[REGIONAL RECRUITMENT (VEG. 1.2)]

fol. 4v Th'electioun of werreours is good — *selection of warriors*
In every londe; and southward ay the more, — *but the further southward*
The more wit thei have and lesse blood,
Forthi to blede thei drede it, and therfore — *Therefore*
Reserve theim to labour and to lore.

2 Lines 149–51: *Our knights take example from them as they can, but truly they are as far from them as gems are from stars*

And northeward hath more blood and lesse
Wit, and to fight and blede an hardinesse.

[Rural Recruits Are Best (Veg. 1.3)]

But werreours to worthe wise and bolde, *become*
Is good to take in mene atwix hem twayne, *common between these two*
Where is not over hote nor over colde. *too hot or too cold*
And to travaile and swete in snow and rayne,[3]
In colde and hete, in wode and feeldys playne, *woods and flat fields*
With rude fode and short, thei that beth used,
To chere it is. The citesens seclused.[4]

And of necessitee, if thei be take *they [city-dwellers] are assigned*
To that honour as to be werreourys,
In grete travaile her sleuth is off to shake. *labor their sloth*
And tolleraunce of sonne and dust and shourys, *sun; showers*
To bere and drawe, and dayes delve and hourys[5]
First use thei, and reste hem in a cave, *First they must learn*
And throute among, and fode a smal to have. *Or outside of it*

In soden case emergent, hem elonge
Fro their cité, streyt out of that pleasaunce.[6]
So shal thei worthe, ye, bothe bolde and stronge. *become, yes*
But feithfully the feld may most avaunce *[from] the country; advance*
A myghti ooste. Of deth is his doubtaunce *army. A man fears his death*
Ful smal, that hath had smal felicité. *Very little, who; little fortune*
To lyve, and lande-men such lyvers be. *rural men*

[Age of Recruits (Veg. 1.4)]

fol. 5r Of yonge folk is best electioun, *selection*
In puberté thing lightlier is lerned, *youth things are easier learned*
Of tendre age up goth perfectioun *From; raised up*
Of chivalers, as it is wel governed. *knights*
Alacrité to lepe and renne unwerned, *leap and run without limits*

3 *And [those who are able] to labor and sweat in snow and rain*

4 Lines 185–86: *With simple and sparse food, those who are used to it, / It brings cheer. [But] city-dwellers are excluded [from that cheer]*

5 *[How] to carry and pull [burdens], and [spend] days and hours digging*

6 Lines 194–95: *When an abrupt situation emerges, hold them / Outside of their city, far away from their comforts*

Not oonly be, but therto sette hem stronge — *set them to it aggressively*
And chere theim therwith, whil thei beth yonge. — *cheer; young*

For better is that yonge men compleyne
On yerys yet commyng and nat fulfilled, — *[fighting] years not yet reached*
Then olde men dolorouxly disdeyne, — *Than; sadly regret*
That thei here yougthe in negligence ha'spilde. — *their youth; have spent*
The yonge may seen alle his daies filde — *days filled*
In disciplyne of were and exercise, — *training for war*
That age may not have in eny wise. — *any way*

Not litil is the discipline of werre, — *No trivial thing; war*
O fote, on hors, with sword or shild or spere, — *On foot*
The place and poort to kepe and not to erre, — *battle station; appropriate demeanor*
Ne truble make, and his shot wel bewere,
To dike and voyde a dike, and entir there, — *dig defensive trenches; excavate*
As is to do; lerned this governaunce,
No fere is it to fight, but pleasaunce.

[Height of Recruits (Veg. 1.5)]

The semelyest, sixe foote or litil lesse, — *most fitting*
The first arayes of the legyoun, — *lines; legion*
Or wyngys horsyd, it is in to dresse. — *wings of horsemen; to take up a position*
Yet is it founde in every regioun,
That smale men have had myght and renoun: — *strength and renown*
Lo, Tideus, as telleth swete Homere, — *Tydeus*
That litil man in vigour had no pere. — *peer*

[Recognizing Good Recruits (Veg. 1.6)]

fol. 5v And him, that is to chese, it is to se — *choose; observe*
The look, the visagynge, the lymys stronge, — *eyes, face, strong limbs*
That thei be sette to force and firmytee;
For bellatours, men, horsis, hondys yonge, — *warriors; hounds*
As thei be wel fetured, is to fonge. — *accept*
As in his book seith of the bee Virgile, — *As Vergil says of the bee in his book*
Too kyndis are, a gentil and a vile. — *noble; worker*

The gentil is smal, rutilaunt, glad-chered, — *fit, ruddy-colored; with a good mood*
That other horribil, elenge, and sloggy, — *wretched, and slothful*
Drawinge his wombe abrede, and ugly-hered, — *Dragging his belly; ugly-haired*
To grete the bolk, and tremulent and droggy, — *Too large his bulk; trembling; sluggish*

The lymes hery, scabious, and ruggy. *limbs hairy; unshapely*
That be wil litil do, but slepe and ete,
And al devoure, as gentil bees gete. *devour all, as noble bees get it*

So for bataile adolescentys yonge
Of grym visage and look pervigilaunt, *ever-alert*
Upright-necked, brod-brested, boned stronge. *broad-chested*
Brawny, bigge armes, fyngeres elongaunt, *long fingers*
Kne deep, smal wombe, and leggys valiaunt *Knees shapely, firm-bellied; ready*
To renne and lepe: of these and suche signys *run and leap; signs*
Th'electioun to make ascribed digne is. *worthy*

[THE TRADES OF GOOD RECRUITS (VEG. 1.7)]

For better is, of myghti werryourys
To have ynogh, then over mych of grete. *enough, rather than too much size*
What crafty men t'abide on werrys shourys, *trades hinder men from battle-storms*
It is to se; fisshers, foulers, forlete *fishermen, fowlers, bypassed*
Hem alle, and pigmentaryes be foryete, *dyers be disregarded*
And alle they that are of idil craftys: *idle trades*
Their insolence and feet to be forlafte is.[7]

fol. 6r The ferrour and the smyth, the carpenter, *ironworker; blacksmith*
The huntere of the hert and of the boor, *deer; boar*
The bocher and his man, bed hem com nere, *butcher; bid them come near*
For alle tho may do and kepe stoor. *maintain merit*
An old proverbe is it: Stoor is not soor, *Merit is not hardship*
And commyn wele it is, a werreour *a common truth*
To have as wel good crafte as grete vigour.

The reaumys myght, the famys fundament,[8]
Stont in the first examynatioun *Depends upon*
Or choys, wheryn is good be diligent. *choice*
Of the provynce that is defensioun; *[its best] defense*
A wysdom and a just intensioun
Is him to have, an ost that is to chese,
Wheryn is al to wynne or al to lese. *In this is everything won or lost*

If chivalers, a land that shal defende, *who shall defend that land*
Be noble born, and have lond and fee, *landed and movable property*
With thewys goode, as can no man amende, *good morals*

---

7 *Their insolence and [their] feet (i.e., their service in the army) are to be rejected*

8 *The realm's strength, the foundation [of its] renown*

Theі wil remembir ay their honesté,
And shame wil refreyne hem not to fle; *will prevent them from fleeing*
Laude and honour, hem sporynge on victory, *Praise; spurring them on [to]*
To make fame eternal in memory.

What helpeth it, if ignobilitee *unworthy [men]*
Have exercise in werre and wagys large? *compensations*
A traitour or a coward if he be,
Thenne his abode is a disceypt and charge. *deceit and burden*
If cowardise hym bere away by barge *bears him away*
Or ship or hors, alway he wil entende *intend*
To marre tho that wolde make or mende.[9]

fol. 6v Civilians or officers to make *Jurists*
Of hem that have habilité to werre, *affinity to combat*
Is not the worship of a lond t'awake. *Will; honor; land awaken*
Sumtyme also lest noughti shuld com nerre;[10]
Thei sette hym to bataile, and theryn erre. *send them [the unworthy] into*
Therfore it is by good discretioun
And grete men to make electioun. *And [by]; to select [recruits]*

[Training: The Will and the Mind (Veg. 1.8)]

And not anoon to knyghthode is to lyft *immediately into knighthood*
A bacheler elect; let first appare *selected recruit; be revealed*
And preve it wel that he be stronge and swift
And wil the discipline of werrys lere, *will learn*
With confidence in conflict as he were.
Ful oftyn he that is right personabil,
Is aftir preef reported right unabil. *demonstration; incompetent*

He putte apart, putte in his place another. *[If] he is pushed back*
Conflicte is not so sure in multitude *not so much about numbers*
As in the myght. Thus proved oon and other
Of werre an entré or similitude, *an introduction to war or its likeness*
Is hem to shewe. But this crafte dissuetude
Hath take away; here is noon exercise[11]
Of disciplyne, as whilom was the gise. *as once was the case*

[9] *To mar [the reputations of] those who would achieve or do even better*

[10] *Lest unworthy men sometimes will come into the army*

[11] Lines 303–04: *But disuse has taken this ability away*

How may I lerne of hym that is unlerned?
How may a thing informal fourme me? 	*instruct*
Thus I suppose is best to be governed: 	*In this way*
Rede up th'istories of auctoritee, 	*the histories of the authorities*
And how thei faught, in theym it is to se,
Or better thus: Celsus Cornelius
Be red, or Caton or Vegetius. 	*Cato [the Elder]*

fol. 7r

Vegetius it is, that I entende
Aftir to goon in lore of exercise,
Besechinge hem that fynde a faut, amende 	*those who; fault [in my work]*
It to the best, or me t'amende it wise: 	*advise*
As redy wil I be with my servyce
T'amende that, as ferther to procede. 	*To amend that (i.e., my own sources)*
Now wel to go, the good angel us lede.

[Training: The Military Pace (Veg. 1.9)]

First is to lerne a chivalerys pace. 	*knight's pace [of steps]*
That is to serve in journey and bataile. 	*marching and battle*
Gret peril is if they theryn difface 	*deface this [good order]*
That seyn our enemye wil our oste assaile, 	*Who say*
And jumpe light; to goon is gret availe, 	*And [thus] jump [too] quickly*
And XX M^L pace in howrys fyve 	*a pace of 20 miles in five hours*
Wel may they goon, and not goon over blyve. 	*too rapidly*

And wightly may thei go IIII M^L moo, 	*(i.e., in forced march) four miles further*
But faster and they passe, it is to renne. 	*if they go faster; run*
In rennyng exercise is good also, 	*running*
To smyte first in fight, and also whenne 	*strike first*
To take a place our foomen wil forrenne[12]
And take it erst; also to serche or sture, 	*first; scout or take action [in battle]*
Lightly to come and go, rennynge is sure.

Rennynge is also right good at the chace, 	*chasing [of enemies]*
And forto lepe a dike is also good, 	*leap [over] a defensive trench*
To renne and lepe and ley uppon the face,
That it suppose a myghti man go wood
And lose his hert withoute sheding blood.[13]

---

[12] *To seize a [favorable] position our foes [try to] run ahead*

[13] Lines 336–38: *[As is] to run and leap and rain [blows] upon the [enemy's] front / That it might make a mighty man go mad [with fear] / And lose his will [to fight] without shedding blood*

For myghtily what man may renne and lepe,
May wel devicte and saf his party kepe. *defeat [his enemy] and keep safe*

[TRAINING: SWIMMING (VEG. 1.10)]

fol. 7v To swymme is eek to lerne in somer season: *swim is also [best] learned in summer*
Men fynde not a brigge as ofte as flood. *find bridges less often than rivers*
Swymmyng to voide and chace an oste wil eson;[14]
Eeke aftir rayn the ryveres goth wood. *Also; rivers run wild*
That every man in th'oost con swymme is good:
Knyght, squyer, footman, cook, and cosynere *squire; kitchen-man*
And grome and page in swymmyng is to lere. *groom; learn*

[TRAINING: SHIELDS AND POSTS (VEG. 1.11)]

Of fight the disciplyne and exercise
Was this: to have a pale or pile uppight *pole or pillar erected*
Of mannys hight, thus writeth olde wyse. *a man's height*
Therwith a bacheler or a yong knyght *recruit*
Shal first be taught to stonde and lerne fight:
A fanne of doubil wight tak him his shelde,[15]
Of doubil wight a mace of tre to welde. *double weight; wood to wield*

This fanne and mace, which either doubil wight is *each of which is double weight*
Of shelde and sword in conflicte or bataile, *[used] in fighting or battle*
Shal exercise as wel swordmen as knyghtys.
And noo man (as thei seyn) is seyn prevaile *say; is said [to] prevail*
In felde or in gravel though he assaile, *In fields or in rocks*
That with the pile nath first gret exercise — *pole did not*
Thus writeth werreourys olde and wise.

Have uche his pile or pale upfixed faste, *each [trainee]*
And, as in werre uppon his mortal foo, *battle; foe*
With wightynesse and wepon most he caste *speed; must he prepare*
To fighte stronge, that he ne shape him fro, *stoutly; not retreat himself from*
On him with shild and sword avised so,
That thou be cloos, and prest thi foo to smyte, *ready*
Lest of thin owne deth thou be to wite. *your own death you will be blamed*

---

[14] *Swimming will facilitate an army in retreat or in chasing down*

[15] *A wicker fan of double [typical] weight he will take for his shield*

fol. 8r Empeche his hed, his face, have at his gorge, *Assault; throat*
Bere at the breste, or serve him on the side *Thrust; breast*
With myghti knyghtly poort, eve as Seynt George, *show of strength; just as St. George*
Lepe o thi foo, loke if he dar abide. *Leap upon your foe*
Wil he nat fle, wounde him, mak woundis wide, *If he will not flee*
Hew of his honde, his legge, his thegh, his armys. *hand; arms*
It is the Turk: though he be sleyn, noon harm is.

[Training: Strike with the Point (Veg. 1.12)]

And forto foyne is better then to smyte. *thrust; than to slash*
The smyter is deluded mony oonys: *The slasher; [on] several counts*
The sword may nat throgh steel and bonys bite, *bones*
Th'entrailys ar covert in steel and bonys. *are covered*
But with a foyn anoon thi foo fordoon is: *thrust; foe is undone*
Tweyne unchys entirfoyned hurteth more *Two inches thrust into [the body]*
Then kerf or ege, although it wounde sore. *blade or edge*

Eek in the kerf, thi right arm is disclosed, *Also in the act of slashing; exposed*
Also thi side; and in the foyn, covert *but in the thrust, covered*
Is side and arm, and er thou be supposed *even before you are*
Redy to fight, the foyn is at his hert *the thrust is [already]*
Or ellys where — a foyn is ever smert. *always fast*
Thus better is to foyne then to kerve; *thrust than to slash*
In tyme and place ereither is t'observe.

This fanne and mace ar ay of doubil wight, *are always of double weight*
That when the bacheler hath exercise *recruit*
Of hevy gere, and aftir taketh light *With heavy gear*
Herneys, as sheeld and sword of just assise, *Harness; standard sizing*
His hert avaunceth, hardynes t'arise. *heart cheers, [his] boldness rises*
"My borthon is delyvered," thinketh he, *burden is lessened*
And on he goth, as glad as he may be.

[Training: Importance of Drills (Veg. 1.13)]

fol. 8v And over this al, exercise in armys
The doctour is to teche and discipline: *drillmaster*
For double wage a wurthi man of armys *double wages*
Was wont to take, if he wer preved digne *able; were proven proficient*
Aforn his prince, ye, tymes VIII or IX. *Before; 8 or 9 times*

And whete he had, and barly had the knyght
That couthe nat as he in armys fight.[16]

Res publica right commendabil is, — *'The common good'*
If chivalers and armys there abounde,
For, they present, may nothing fare amys, — *[where] they [are] present; amiss*
And ther thei are absent, al goth to grounde. — *all goes to dirt*
In gemme, in gold, in silk be thei fecounde, — *gems; even if they are numerous*
It fereth not. But myghti men in armys — *Cause no fear [among enemies]*
They fereth with the drede of deth and harmys. — *harms*

Caton the wise seith: where as men erre — *Cato; says: whenever*
In other thinge, it may be wel amended;
But emendatioun is noon in werre: — *none in war*
The cryme doon, forthwith the grace is spended,[17]
Or slayn anoon is he that there offended,
Or putte to flight, and ever aftir he
Is lesse worth then they that made him fle.

[TRAINING: THROWING SPEARS (VEG. 1.14)]

But turne ayeyn, inwit, to thi preceptys! — *again, mind, to your teachings*
With sword and sheld the lerned chivaler
At pale or pile, in *artilaunce* excepte is;[18]
A dart of more wight then is mester, — *javelin; weight; standard*
Tak him in honde, and teche hym it to ster, — *steer*
And caste it at that pile, as at his foo,
So that it route, and right uppon hym go. — *root [into it]*

fol. 9r Of armys is the doctour heer t'attende,
That myghtily this dart be take and shake,
And shot as myghtily, forthright on ende,
And smyte sore, or nygh, this pile or stake; — *strike deep, or near*
Herof vigour in th'armys wil awake
And craft to caste and smyte shal encrece;
The werreours thus taught, shal make peax. — *peace*

[16] Lines 402–03: *And wheat he [who was proven proficient] had, while the knight had barley / Who could not fight as well in arms as he did*

[17] *The error [being] done, at once the grace is squandered (i.e., any chance for forgiveness is suspended)*

[18] Lines 419–20: *The knight experienced with sword and shield / At the pole or post, accepted is into the art of spear-fighting*

[Training: Archery (Veg. 1.15)]

But bachilers, the thridde or firthe part,	*a third or quarter part of recruits*
Applied ar to shote in bowes longe	*Are trained to shoot*
With arowys. Heryn is doctryne and art:	*doctrine*
The stringys up to breke in bowes stronge,[19]	
And swift and craftily the taclis fonge,	*let fly the arrows*
Starkly the lifte arm holde with the bowe,	*Firmly hold the left arm*
Drawe with the right, and smyte and overthrowe.	
Set hert and eye uppon that pile or pale,	*Focus the mind and eye*
Shoot nygh or on, and if so be thou ride,	
On hors is eek the bowys bigge uphale.[20]	
Smyte in the face or breste or bak or side,	
Compelle fle, or falle, if that he bide.	*flight; remain*
Cotidian be mad this exercise,	*Daily make this practice*
On fote and hors, as writeth olde wise.	*as old authorities say*
That archery is grete utilitee,	
It nedeth not to telle eny that here is.	*any*
Caton, therof in bookys writeth he,	*Cato*
Among the discipline of chivalerys,	*In [his books on]*
And Claudius, that werred mony yeres,	*fought for many years*
Wel seide, and Affricanus Scipio	*Scipio Africanus*
With archerys confounded ofte his foo.	*foe*

[Training: Casting Stones (Veg. 1.16)]

fol. 9v Use eek the cast of stoon with slynge or honde.	*casting of stones*
It falleth ofte, if other shot ther noon is,	*It often happens*
Men herneysed in steel may not withstonde	
The multitude and myghti caste of stonys.	
It breketh ofte and breseth flesh and bonys,	*bruises*
And stonys in effecte are everywhere,	
And slyngys ar not noyous forto bere.	*troublesome to carry*
And otherwhile in stony stede is fight,	*often warfare happens in stony places*
A mountayn otherwhile is to defende,	*[Or]*
An hil, a toun, a tour, and every knyght	*tower*
And other wight may caste stoon on ende —	*person*
The stonys axe, if other shot be spende.	*Ask for the stones*

[19] *[How] to string up strong bows to the point of breaking*

[20] Lines 440–41: *Shoot near or into [them]; and if it happens that you ride, / On horseback bows are raised*

Or ellys thus: save other shot with stonys,
Or use hem, as requireth, both at onys.

[TRAINING: THROWING DARTS (VEG. 1.17)]

The *barbulys* that named ar plumbatys, *'lead balls'; missiles*
Set in the sheld is good to take fyve.
That used hem of old wer grete estatys.[21]
As archerys, they wolde shote and dryve
Her foo to flight, or leve him not alyve. *Their foe*
This shot commended Dioclisian *Diocletian*
And his co-emperour, Maxymyan. *Maximian*

[TRAINING: MOUNTING HORSES (VEG. 1.18)]

The chivalers and werreourys alle, *knights*
Quicly to lepe on hors, and so descende
Uppon the right or lyft side, if it falle, *however it happens*
That exercise is for to kepe an ende:
Unarmed first, and armed thenne ascende,
And aftir with a spere or sword and shelde.
This feet is good when troubled is the felde. *feat*

[TRAINING: CARRYING GEAR (VEG. 1.19)]

fol. 10r And LX pounde of weght is good to bere, *60 pounds of weight; bear*
And go therwith a chivalerys pace, *knight's pace (i.e., a military step)*
Vitaile and herneysing and sword and spere, *Victuals and harnessing*
Frely to bere; al this is but solace. *Readily to bear; child's play (ironic)*
Thinge exercised ofte in tyme and space, *Things practiced*
Hard if it be, with use it wil ben eased, *will get easier*
The yonge men herwith beth best appesed.[22]

[TRAINING: WEARING ARMOR (1.20)]

And exercise him uche in his armure, *armor*
As is the gise adayes now to were. *manner nowadays to wear*
And se that every peece herneys be sure. *piece of harness is secure*

---

[21] *Those who used them in old days were great countries*

[22] *The young men (i.e., inexperienced recruits) will be best placated [by knowing this]*

Go quycly in, and quyk out of gere,
And kepe it cler, as gold or gemme it were. *clean, as [if] it was [made of]*
Corraged is that hath his herneys bright, *Braver is [he] who; harness*
And he that is wel armed, dar wel fight.

[Training: Keeping Watch (Veg. 1.21)]

To warde and wacche an oste it is to lerne *guard and watch an army; learn*
Both holsom is that feet and necessary. *proper; skill*
Withinne a pale an oste is to governe, *palisade*
That day and nyght safly theryn they tary *safely they remain*
And take reste, and never oon myscary. *have no one come to harm*
For faute of wacch, ha worthi not myscheved
Now late, and al to rathe? Is this nat preved?[23]

[Where to Construct a Camp (Veg. 1.22)]

To make a fortresse, if the foon be nygh, *fort; foes are near*
As sure a grounde, and se that ther be fode *see that there is food*
For man and beest, and watir deep myd-thigh, *[at least] mid-thigh deep*
Not fer; and se there wode or grovys goode. *Not far [away]; wood and trees*
Now signe it, lyne it out by yerde or rode. *allocate it, measure; yard or rod*
An hil if ther be nygh, wherby the foo
May hurte, anoon set of the ground therfro.[24]

fol. 10v Ther flood is wont the felde to over-flete, *Where [a]; to overflow the ground*
Mak ther noo strength; and as is necessary *no stronghold*
Unto thyn oste, as mych is out to mete — *For your army, that much measure out*
And cariage also theryn most tary. *the baggage-train*
Men dissipat, here enemy may myscary, *spread out, their; harm [them]*
And combred is an oste that is compressed; *obstructed*
Tak eve ynough, and hoom have uch man dressed.[25]

[Shape of a Camp (Veg. 1.23)]

Trianguler, or square, or dymy-rounde, *semi-circular*
The strength it is to make of hosteyinge, *for [each] campaign*

[23] Lines 500–01: *If not for the failure of the watch, would the nobles have been harmed / Recently, and all at once? Is this not proven?*

[24] Lines 507–08: *If there is a hill nearby, from which a foe / May harm [you], mark your position far enough away*

[25] *Take enough [space] nightly, and have each man assigned to lodgings*

Th'avis therof is taken at the grounde.	*plan*
And estward, or uppon thi foo comynge,	*in the direction of the foe*
The yatys principal have ussuynge,	*primary gates for issuing out*
To welcom him; and if an ost journey,	*army is on the march*
Tho yatis ar to sette uppon his wey.	*Those gates*
The centenaryes theruppon shal picche	*centuries (i.e., the cohorts); pitch*
Her pavilons, and dragonys and signys	*Their; dragons and standards*
Shal up be set, and Gorgona the wicche	*[images of] a Gorgon; witch*
Up sette they — to juste batail condigne is	*permitted*
Uch helply thing. Another yate and signe is:	*[there] is*
Ther trespassers shal go to their juesse,	*judgement*
That oponeth north, or westward, as I gesse.	*opens*

[CONSTRUCTING A CAMP (VEG. 1.24)]

In III maneer a strengthe is to be walled,	*By three means a fort*
If ther oppresse noo necessitee.	
Delve up the torf, have it togedir malled,	*turf; piled*
Therof the wal be mad high footys III	*three feet*
Above grounde; the dike withouten be	*ditch outside*
IX foote brode, and VII deep dounright;	*nine feet broad; seven [feet]*
Thus dike and wal is wel X fote in hight.	
fol. 11r This werk they calle a dike tumultuary.	*temporary (hastily raised) ditch*
To stynte a rore, and if the foo be kene,	*stop a tumult; foe is keen*
Legytymat dykinge is necessary:	*Proper ditchwork (i.e., a fosse)*
XII foote brod that dike is to demene,	*measure*
And nyne deep, his sidys to sustene.	*to sustain (i.e., using revetments)*
And hege it as is best on either side,	*lined [with revetments]*
That diked erth upheged stonde and bide.	*dug-up earth heaved up*
Above grounde arise it foure foote:	*they (i.e., the revetments) arise*
Thus hath the dike in brede footys XII,	
And XIII is it high fro crop to roote.	*(i.e., from top to bottom)*
That stake of pith which every man himselve	*stake of strong wood*
Hath born, on oneward is it forto delve.	*Have carried, are driven in above*
And this to do, pikens, mattok, and spade	*picks, mattocks*
And tole ynough ther most be redy made.	*tools; must be made ready*

[Constructing a Camp Under Threat (Veg. 1.25)]

But and the foo lene on forwith to fight,	*But if the foe moves nearby*
The hors men alle, and half the folk o'foote	*horsemen; footmen*
Embataile hem, to showve away their myght.	*Engage them*
That other half, to dike foot by foote,	*dig the ditch*
Be sette, and an heraude, expert by roote,	*herald, according to form*
The centrions other the centenaryis	*centurions; centuries*
In ordre forth hem calle, as necessary is.	
And ay among the centrions enserch,	*ever among [them] the centurions inspect*
The werk, if it be wrought, kept the mesure,	
In brede and deep and high, perch aftir perch,	*bit by bit*
And chastise him that hath nat doon his cure.	*punish; not done his duty*
An hoste thus exercised may ensure	*army thus trained; trust*
In prevalence, whos debellatioun	*In victory; defeat*
Shal not be straught by perturbatioun.	*caused by disarray*

[Training: Maintaining Order in the Lines (Veg. 1.26)]

fol. 11v Wel knowen is, nothinge is more in fight	*known [it] is; more [advantageous]*
Then exercise and daily frequentaunce.	*practice*
Uch werreour therfore do his myght	*Each; does*
To knowe it wel and kepe his ordynaunce:	*know and maintain his order well*
An ooste to thicke, I sette, is encombraunce,	*An army too dense, I declare*
And also perilous is over-thynne,	
Thei sone fle that be to fer atwynne.	*too far between*
We werreours, forthi go we to feelde;	
And as our name in ordir in the rolle is,	
Our ordunaunt so sette us, dart and sheelde	*ordinarius*
And bowe and axe, and calle us first by pollys.	*individually [as units]*
Triangulys, quadrangulys, and rollys,	*Triangles, rectangles; circles*
We may be made; and thus us embataile,	*made [to form]*
Governed, undir grate to prevaile.	*under attack*
A sengil ege is first to strecch in longe,	*single line; length*
Withoute bosomynge or curvature,	*bending in or out*
With dowbeling forwith let make it stronge,	*Then doubling the line; stronger*
That also fele assiste, in like mesure.	
And with a woord turne hem to quadrature,	*rectangle [formation]*
And efte trianguler, and then hem rounde,	
And raunge hem efte, and keep everych his grounde.	*range; in his order*

This ordynaunce of right is to prevaile.
  Doctryne hem eek whenne it is best to square,   *Teach them also*
And when a triangul may more availe,
  And orbys, how they necessary are,
How may be to condense, and how to rare.   *how to tighten up; disperse*
  The werreours that ha this exercise,   *have*
  Be preste with hardynesse, and stronge and wise.   *instilled*

[Marching Maneuvers (Veg. 1.27)]

fol. 12r And over this, an olde usage it was   *practice*
  To make walk thryes in every mone,   *march thrice a month*
And tho they wente a chivalerys paas   *[at] a knight's pace (i.e., military step)*
  X myle outward, the men of armys, none   *Ten miles*
Unharneysed: the footmen everychone
  Bowed, tacled, darted, jacked, saladed;   *(see note)*
  Vitaile eke born withal, her hertis gladed.   *Victuals also carried; hearts*

In hom comynge among thei wente faste   *Returning to camp at times*
  And ranne among. Eek tourmys of ryderys   *ran; troops of riders*
Sumtyme journeyed on foote in haste,
  Shelded and herneysed with myghti sperys —
Not oonly in the playn, but also where is
  A mountayn or a clif or streyt passagys.   *difficult paths*
  Thus hadde thei both exercise and wagys.

Ereithre ege in this wise exercised   *All lines [of the army]; way*
  Was by and by, so that no chaunce of newe   *surprise*
Nas to be thought that thei nere of avised,[26]
  And hadde way the daungerys t'eschewe   *to avoid*
Undaungered. And this wisdom thei knewe   *Unharmed*
  By discipline of their doctour of armys,   *drill instructor*
  To wynne honour withouten hate or harmys.   *injury or wounds*

[End of Part I]

Th'electioun and exercise anended,   *Recruitment and training complete*
  An ooste is now to numbre and dyvide   *ordered in numbers and divisions*
And seen uch officer his part commended,
  And how to sette a feeld to fight and bide.   *fight and engage*
Goode angelys and sayntys, ye me gide

[26] *Could be thought of that they had not prepared for*

And lighte me, O Lady Saynte Mary!
To write wel this werk and not to tary. *tarry*

Part II: The Military Structure

[Proem: In Praise of the King]

fol. 12v *Electrix ita milicie pars prima recedit,*
Et pars partitrix ecce secunda subit.[1]

The firste part of IIII is here at ende; — *of 4 parts*
Now to the part secounde! Er we procede — *Before*
To knowe this, His grace God us sende!
Myn auctour ofte adviseth us to rede — *My author often advises; read*
And to the sense of it to taken hede; — *heed*
To rede a thinge withoute intelligence, — *understanding*
As seith Cato the wise, is negligence.

But this I leve unto the sapience — *wisdom*
Of chivalers, and to my werk retorne, — *knights*
Theryn to do my feithful diligence
For their pleasaunce, out of this prosis storne — *stern prose [Vegetius's treatise]*
The resonaunce of metris wolde I borne. — *rhyme would I birth*
As myghti herte in ryngynge herneysinge,
So gentil wit wil in good metris springe.[2]

And for th'onour of th'everlastyng Kynge,
Our Saviour Jhesus, and His ierarkys, — *hierarchs*
His angelys, and for that swete thinge,
His Modre, patronesse of al my warkys, — *Mother; all my works*
For His prophetys love and patriarkys, — *and [that of His] patriarchs*
And for th'Apostolis that made our Crede, — *Creed*
As do me favour, ye that wil me rede. — *you who will read my words*

---

1 Before 621: *The first part, on the selection of military men, thus withdraws, / And behold, the second part, on divisions, enters*

2 Lines 633–34: *Just as a mighty heart [will do so] in [hearing] a ringing harness, / So will a noble mind be thrilled with good poetry*

[Divisions of Armed Forces (Veg. 2.1)]

fol. 13r Virgile seith (an high poete is he!) — *lofty*
That werre in armys stont and mannys myght:[3]
The man on hors, o fote, or on the see. — *horse, on foot; sea*
Riders be wyngis clept, for swift and light, — *are called wings*
On either half of th'ege eke ar thei dight. — *either side of the [front] line; found*
But now that ege is called the banere — *line; banner*
Or banerye, havyng his banereer. — *bannerie; bannerer (i.e., standard-bearer)*

Also ther are riders legyonaryis; — *legionary riders*
Thei are annexed to the legioun.
In too maner of shippes men to cary is, — *two kinds of ships men are carried*
Their namys ar couth in this regioun: — *known*
Or-thwart go they the flood, and up and doun. — *Crossing over; waters*
Riders in playn, footmen goth everywhere, — *[go] in plains, footmen go*
By theyme the commyn wele is to conquere. — *common profit*

Riders a fewe, and have o foote fele,
Thei spende smal, and horsmen spende fre.[4]
Footmen o tweyne is to dyvide and dele: — *are divided in two*
Or legiaunt or aydaunt for to be. — *Either legionary or auxiliary*
Confederat men aydaunt is to se, — *Allied men are to be seen as auxiliaries*
That is to say, by trewce or toleraunce, — *truce*
As Frensh ar suffred here, and we in Fraunce. — *French are suffered [to be]*

Aydaunt be they, but in the legioun — *Auxiliary*
Lith th'ordinaunce in werre to prevaile. — *Lies the appointed duty; war*
A legioun out of electioun — *selection*
Hath take his name, as elect to bataile: — *Takes its name; selected to do battle*
Her diligence and feith is not to faile. — *Their; faith; fail*
Thi legyaunt forthi to multiplie
Is right, but aydauntys a fewe applie.[5]

[Comparing the Legion and the Auxiliary (Veg. 2.2)]

fol. 13v VIII thousant werreours was a phalange — *8,000 warriors were a phalanx*
In dayis olde, and of VI M1 men — *6,000*

---

3 *That war consists of arms and of men's strength*

4 Lines 656–57: *Riders [have] a smaller number, and footmen have far more, / [Because] they (i.e., footmen) cost little, and horsemen cost a lot*

5 Lines 668–69: *For the legion to have a much larger number [of men] / Is right, with relatively fewer men in the auxiliary*

Was a caterve, but this diagalange *caterva (i.e., horde); such detail*
Is, as to this, not worth a pulled hen.
The legioun, departed into X, *divided into 10 parts*
Is us to lerne, and legions how fele *learn*
It is to have, and how asondir dele. *asunder*

[Formation of the Legion (Veg. 2.4–5)]

The consules II legiounys ladden, *consuls [each] led two legions*
Al aldermeest; but thei hadde exercise, *At most; training*
Wherof the felde victoriously thei hadden. *battlefield*
To chose a legioun, this was the gise, *advice*
In bookys as they seyn, these olde wise: *authorities*
Wyis, hardy, strong, doctryned, high statured, *Wise; obedient*
In feet of werre ofte used and wel ured. *feats of war (i.e., feats of arms); trained*

That was the man, he was mad mylitaunt,
When al the world to the Romayn Empire
Was made obey, by knyghthod valiaunt. *made [to] obey*
A sacramental oth doth it requyre, *oath*
To write pleyn this matere I desire,
By God and Criste and Holy Goost swar he, *he swore*
And by that emperourys majestee. *emperor's*

Next God is hym to drede and hym to honoure is.
Right as to God ther bodily, present
To th'emperour, when he mad emperour is,
Devotioun — uch loyal ympendent
Is to be vigilaunt, his servyent.[6]
God serveth he, both knyght and comynere, *commoner*
That loveth hym, to God that regneth here. *reigns*

fol. 14r God, Criste Jhesus, and Holy Goste was sworn *were sworn [to]*
By theim, and th'emperourys majestee,
That his commaundementys shuld be born *orders should be borne*
And strenuously be doon, be what thei be. *done*
Fro mylitaunce that thei shal never fle *military service; desert*
Ner voyde deth, but rather deth desire *Nor avoid death*
For th'emperour, and wele of his Empire. *well-being*

[6] Lines 691–95: *Second [only to] God [it] is [right] to fear and honor him (i.e., the emperor). / As rightly as if God [is] there bodily, present / To the emperor, when he is made emperor, / Devotion — [for] each loyal subject / Is to be vigilantly his servant*

[COHORTS AND SOLDIERS (VEG. 2.6)]

Thus sworn, uch knyght is of the legioun.
The legioun stont in cohortys X; *10 cohorts*
Cohors the Latyn is, this regioun
T'English it fore, help us, good Lord! Amen. *To [translate] it into English*
The dignité and number of the men
Hath in the firste cohors an excellence *first cohort*
Of noble blood, manhode, and sapience. *wisdom*

This feleshepe, most worshipful, most digne,
Bar th'egil and th'ymage of th'emperour; *Carried the eagle*
As God present was holden either signe,
Thei hadde both attendaunce and honour;
Of chivalers heryn was doon the flour, *flower (i.e., the best)*
A M^{1} an C and V footmen, *1,105*
And VIII^{XX} and VI of wight horsmen. *166 strong*

The military cohors, or the choors, *military cohort*
Thus named it the wise. And the secounde
Cohors, like as the bonet to his coors *bonnet to the body*
Is set, thei sette it footmen stronge and sounde,
V C and an half, and V abounde *555*
In hit, with sixe and sixti hors, and it *66*
The quyngentary called men of wit. *quingenary [cohort]*

fol. 14v As fele and myghty choys putte in the thridde is, *strong; selection [of men] is put*
For in their honde espoyre is al to thryve: *hands hope*
Her place in ordynaunce is in the myddys. *Their; middle*
And for the firth choors is to discrive *fourth cohort; assign*
Footmen V C and an half and V, *555*
With sixe and sixti hors, and eve as fele, *even as many*
With better hors, unto the fifthe dele. *horsemen; part*

For as the first cohors is the right horn, *horn (i.e., wing)*
So in the lift horn is the fifthe choors; *left; cohort*
For V choors stonde in the frounte aforn, *5 cohorts*
Or the vawarde: of termys is noo foors, *vanguard; terms are of no concern*
So the conceyt be had. The sixt cohors *If the concept*
Hath, as the fifthe, yet lusty men and yonge;
To th'egil next to stonde it is to fonge. *the eagle; to take up [position]*

That is the right horn. In the myddil warde
The nexte choors hath eve as mony as she, *cohort has as many as it [did]*
The nexte as fele, and therto is tawarde *as strong (i.e., as many)*

The myghti men, amyddis forto be.
The IXth is of the same quantitie, — *ninth [cohort]*
The Xth is eve as is the choors beforn, — *tenth [cohort]*
But make it strong, for it is the lift horn. — *left wing*

The legioun in X is thus cohorted,
VI M^{l} and an C men on foote, — *6,100*
VII C hors and XXX therto soorted, — *730*
Of fewer hors is not to speke or moote
In eny legioun; yet, crop and roote
To seyn, of hors ther may be take moo, — *more*
Commaundement if ther be so to do.

[Officers in the Legion Still in Practice (Veg. 2.7)]

fol. 15r Exployed heer th'usage and ordynaunce — *Having explained here*
Of legyoun, unto the principal — *office*
Of chivalers retourne our remembraunce: — *knights*
The dignitie and name in special
Of every prince enrolled, and who shal — *prince (i.e., officer)*
Do what, and whenne, and where, it is to write.
Good angel, help us al this werk t'endite! — *to compose*

The grete trybune is mad by th'emperour, — *tribune; made*
And by patent, and send by jugement. — *[letter] patent; decree*
Th'undir-trybune is hent of his labour. — *subtribune; identified by*
An ordyner for fighters forth present — *ordinarius*
Is forto sette. Eek th'emperour content — *Also*
Is ofte to sende and make secoundaryis; — *secondarius*
What name is heer for hem? Coordinaryis. — *Coordinarius*

An egiller bar th'egil, and th'ymage — *aquilifer carried the eagle; image-standard*
Of th'emperour bar an ymaginary, — *imaginarius*
And moo then oon ther were of those in wage. — *more than one of those*
A banereer, tho clept a draconary,[7]
A kyng heralde, tho clept a tesserary, — *king's herald; tesserarius*
The baner he, he bar commaundement,[8]
Al th'oost t'obeye her princys hole entent. — *The whole army*

Campigeners made exercise in feeldys, — *Campigeni (i.e., drillmasters); training*
Campymeters mesured out the grounde, — *Campimetes (i.e., surveyors)*

7 *A bannerer, those [were] called a draconarius*
8 *The former one [carried a] banner, the latter carried [an] order*

To picche pavilons, tentys, and teeldys, *pitch; shelters*
The forteresse triangeler or rounde *fortress triangular*
Or square to be made or dymyrounde,
His part hit was. And he that was library, *librarius (i.e., clerk)*
Th'accomptys wrot, that rekenyng ne vary. *Wrote the accounts*

fol. 15v The clarioner, trompet, and hornycler, *bugler, trumpeter; horn-blower*
With horn, and trompe of bras, and clarioun,
In terribil batailis bloweth cleer, *battles blow*
That hors and man rejoyceth at the soun. *sound*
The firmament therto making resoun *heavens; echo*
Or resonaunce; thus joyneth thei bataile —
God stonde with the right, that it prevaile!

A mesurer, that is our herbagere, *measurer; quartermaster*
For paviloun and tent assigneth he
The grounde, and seith: "Be ye ther, be ye here!"
Uch hostel eek, in castel and citee, *Each lodging also*
Assigneth he, uch aftir his degré. *in accordance with a man's class*
A wreth o golde is signe of grete estate; *wreath of*
That wered it, was called a 'torquate'. *whoever wore; torquatus*

Sengil ther were of these, and duplicate *Singular*
And triplicate, and so to IIII and V, *4 and 5*
That hadde wage, uche aftir his estate. *[multiplied] wages*
Tho namys goon, such personys alyve,
It may be thought, therof wil I not scryve.[9]
Ther were eek worthymen clept candidate, *called candidati*
And last, the souldeours, uch othrys mate. *each other's comrade*

[Officers in the Legion Now Obsolete (Veg. 2.8)]

The principal prince of the legioun, *foremost princeps*
Sumtyme it was, and yet is a like gise,
To make a 'primypile', a centurioun — *centurio primi pili*
A lieutenaunt men calle him in our wise. *manner*
And him beforn is th'egil forto arise.
IIII C knyghtis eek of valiaunce *400 valiant knights (i.e., of the first line)*
This prymypile hadde in his governaunce.

[9] Lines 799–800: *Those titles no longer used, [though] such persons still live, / It may be reckoned, I will write of them no more*

fol. 16r	He in the frounte of al the legioun	
	Was as a vi-captayn, a governour,	*vice-captain*
	And took availe at uch partitioun.	*advantage of each wage-giving*
	The first spere was next, a lusty flour:	*first spear (primus hastatus); flower*
	CC to governe is his honour,	*200 spears*
	Wherof thei named him a ducennary,	*ducenarius*
	The name fro the numbir not to vary.	
	The prince an C and an half governed,	*princeps; 150 [men]*
	Eek he governed al the legioun	*Also he administrated*
	In ordynaunce; overal he went unwerned.	*unhindered*
	The nexte spere, of name and of renoun,	*second spear (secundus hastatus)*
	As mony hadde in his directioun.	
	The first triari hadde an C men.	*triarius prior (i.e., third-ranked officer); 100 men*
	A chevetayn was eke of every X.	*chieftain (i.e., captain) [there]; for every 10 men*
	Thus hath the first cohors V ordinayris,	*5 officers*
	And every X an hed, a cheveteyne,	
	To rewle theim; and so it necessayr is,	
	An C and V on this choors to reigne:	*105 men; cohort to govern*
	IIII ordinayris and the cheef captayne,	*4 officers*
	That is their ordinary general,	
	And seyde is ofte of him: he rewleth al.	*rules everything*
	So high honour, so gret utilitee	*great privilege*
	Hath everych estate of this renoun	
	Provided hem by sage antiquitee,	*[to] them*
	That every persone in the legioun	
	With al labour, with al devotioun	
	To that honour attended to ascende,	
	And that avail to wynne, her bodyis bende.	*benefit to win, they labor*
fol. 16v	The nexte choors, named the quyngentary,	*cohort; quingenary*
	Hath centurions or centenerys V;	*centuries*
	Thridde choors as fele hath necessary;	*just as many has*
	The firthe V, and, forto spede us blyve,	*to hasten us along*
	In every choors the centyners oo fyve	*there are 5 centuries*
	In numbir make, and so the legioun	
	Of hem hath LV up and doun.	*55 [men] in total*
	Not LX, whi? For V th'ordinayrys	*Why not 60 [men]?; the 5 officers*
	In their estate and stede of V stonde.	*[have] 5 who can stand for them*
	To graunte this, me semeth, noo contrary is:	*it seems to me*
	Though in my book so wryton I ne fonde,	*I do not find it written*
	Of LV, wel I undirstonde	*From [the total of] 55 [men]*

And fynde cleer, so that it most appere,
That undir-ordynayrys V were. *[there] were 5 subordinarii*

[Prefect of the Legion (Veg 2.9)]

The consulys, for th'emperour, legatys *legates*
Sende unto the oste; to thaim obtemperaunt *army; obedient*
Was al the legioun, and al th'estatys. *the ranks*
They were of al the werres ordynaunt. *wars the administrators*
To theim obeyed everych aydaunt, *auxiliary [force]*
In stede of whom illustres lordes, poerys, *illustrious; powers*
Be substitute, maistrys of chivalerys; *Are [now] substituted, masters*

By whom not oonly legiounys twayn, *not just two legions*
But grete numbrys hadde governaunce. *[even] greater*
The propre juge is the provost, certayn, *commanding officer; prefect*
With worthinesse of the first ordynaunce;
The vi-legate is he by mynystraunce *administration*
Of his power, to hym the centeners *centuries*
Obey, and the trybune and chivalers.

fol. 17r Of him the rolle of wacch and of progresse *password for the nightwatch*
Thei crave and have, and if a knyght offende,
At his precepte he was put to juesse *sent for judgment*
By the trybune, in payne or deth t'anende. *death to bring about*
Hors, herneys, wage, and cloth, vitail to spende, *harness; clothes*
His cure it was t'ordeyn, and disciplyne *duty; to arrange*
Unto every man, severous or benygne. *severe or benign*

His justising, with sobre diligence, *conducting of justice; sober*
And pité doon uppon his legioun,
Assured hem to longh obedience *long*
And reverence, and high devotioun.
Good governaunce at his promotioun
Kept every man; and his honour, him thoughte
It was, when every man dede as him oughte. *did*

[Prefect of the Camp (Veg 2.10)]

The maister or provost of ordynaunce, *ordinance*
Although he were of lower dignitie,
His estimatioun and governaunce,
The bastilys, dich, and pale is to se, *bastions, ditch, and wall*

And wher the tabernaculys shal be — *places of worship*
And tent and teelde and case and paviloun — *shelter; huts*
And cariage of al the legioun. — *baggage*

For seeke men the leche and medycyne — *sick; leech (i.e., healer)*
Procureth he, for larderye and toolys. — *larder; tools*
Of every werk cartyng he most assigne, — *transportation*
For bastile or engyne or myne. And fole is — *[siege] engine or mine; fool*
He noon, that is expert in these scolys: — *none; schools [of study]*
This was a wise, appreved chivalere, — *selected*
That, as he dede himself, couth other lere. — *could teach others*

[Prefect of Engineers (Veg 2.11)]

fol. 17v And over this, the ferrour and the smyth, — *ironworker; blacksmith*
The tymbre men, hewer, and carpenter, — *wood*
The peyntour, and uch other craft goth with, — *painter; every craft [that] goes with it*
To make a frame or engyne everywhere, — *building or [siege] engine*
Hem to defense and her foomen to fere, — *Them; their foes to make afraid*
Tormentys olde and carrys to repare — *Weapons; carts to repair*
And make newe, as they to broken are.

Forgys and artelryis, armeryis, — *Forges; workshops, armories*
To make tole, horshoon, shot, and armurre, — *tools; horseshoes; armor*
And every thing that nede myght aspie is — *might be seen as needed*
In th'ooste; and eek mynours that can go sure — *the army; miners*
Undir the dich, and al the wal demure — *wall undermine*
Or brynge in th'oost; herof the maister smyth — *master blacksmith*
Had al the rule, and ever went he with.

[Tribune of the Soldiers (Veg. 2.12)]

The legioun is seide have choorsis X. — *10 cohorts*
The military first, or miliary, — *milliary*
The best and gentilest and wisest men
And myghtiest, therto be necessary;
Eek letterure is good and light to cary. — *Also learning*
Her governour was a trybune of armys, — *Their; tribune*
Wise and honest, that body strong and arm is. — *armed*

The choorsys aftir that, trybunys cured — *cohorts; tribunes led*
Or maysterys, as it the prince pleased.
Uch chivaleer in exercise assured — *Each; training*

So was, that God and man therwith was pleased.
And first to se the prince do, mych eased — *To see the leader do [things] first*
The hertys alle. Fresh herneys, armur bright, — *hearts; harness*
Wit, hardinesse, and myght had every knyght.

[Lower Officers (Veg. 2.13)]

fol. 18r The firste signe of al the legioun
An egil is, born by an egeler. — *eagle; carried; aquilifer (i.e., eagle-bearer)*
And thenne in every choors is a dragoun, — *cohort*
Born by a draconair or banereer. — *draconarius or bannerer*
A baner eek had every centener — *century*
Other a signe, inscribed so by rowe, — *Or ensign; [a] row [of letters]*
His chevetayn that every man may knowe.

The centeners had also werreourys, — *warriors (i.e., centurions)*
Hardy, wel harneysed, in their salet — *sallet (i.e., light helmet)*
That had a creste of fetherys or lik flourys, — *flowers*
That noon errour were in the batail set, — *no mistakes; formation*
To his cristate and to his baneret — *crested officer*
And to his decanair everych his sight — *decanus (i.e., squad leader)*
May caste, and in his place anoon be pight. — *position at once be fixed*

[Mounted Officers (Veg. 2.14)]

Right as the footmen have a centurion,
That hath in rewle an C men and X, — *110 footmen*
So have the riders a decurion,
That hath in rewle XXXII horsmen. — *32 horsemen*
By his banere him knoweth alle his men,
And over that, right as it is to chese
A myghti man for thaym, so is for these. — *them (i.e., the footmen)*

For theim a stronge and wel-fetured man, — *strong and well-built*
That can a spere, a dart, a sword wel caste, — *wield well*
And also fight, and rounde a sheld wel can, — *manage a shield*
And spende his wepon wel withoute waste, — *use; without wasting effort*
Redier to fight then flite, and ner agaste, — *flee; never scared*
That can be sobre, sadde, and quyk and quyver, — *agile*
And with his foo com off and him delyver;[10]

[10] *And against his foe come off [a blow] and deliver [one against] him*

fol. 18v Obeyssaunt his premynentys wille, *Obedient [to] his commander's*
    And rather do the feat then of it crake, *the [requested] deed; mouth off*
Impatient that day or tyme spille, *be wasted*
    In armys exercise and art to wake, *For armed training*
And of himself a sampeler to make *an example*
    Among his men, wel shod, honestly dight,
    And make hem fourbe her armure ever bright. *polish their armor*

Right so it is, for these men to chese *these [horsemen]; choose*
    A decurioun, thorugh lik to him in fourme, *thoroughly akin*
Impatient that thei the tyme lese, *lose*
    Wel herneysed, and everych of hys tourme *troop*
In every poynt of armys wil enfourme, *[he] will instruct*
    And armed wil his hors so sone ascende, *soon*
    That mervaile is, and course hym stronge anende, *ride strongly at last*

And use wel a dart, a shaft, a spere,
    And teche chivalers undir his cure, *care*
Right as himself to torne hem in her gere, *travel in their gear*
    The brigandyn, helmet, and al procure, *brigandine (light body armor)*
It oftyn wipe clene, and knowe sure, *wipe clean (i.e., polish)*
    With herneysing and myghti poort affrayed *show of strength afraid*
    Is ofte a foo, and forto fight dismayed. *foe*

Is it to sey, "He is a werrely knyght," *warlike*
    Whos herneys is horribil and beduste, *dusty*
Not onys used in a fourte nyght, *once; fortnight*
    And al that iron is or steel, beruste? *rusted*
Unkept his hors, how may he fight or juste? *Ungroomed; joust*
    The knyghtis and her horsys in his tourme *their; troop*
    This capitayn shal procure and refourme. *prepare*

Part III: Field Tactics

fol. 19r *Tercia bellatrix pars est et pacificatrix,*
In qua quosque bonos concomitatur honos.[1]

[Proem: The Present Yorkist Position]

Comprised is in smal this part secounde, — *second part [of the poem]*
An ooste to numbir, and a legioun. — *The numbers in an army*
In foylis is it fewe, in fruyt fecounde: — *folios; fruit fertile*
The salvature of al religioun — *salvation*
Is founde heryn for every regioun.
Wel to digeste this, God graunte us grace, — *understand*
And by the werre His reste to purchace! — *war His peace*

O gracious our kyng! Thei fleth his face. — *They (i.e., the Yorkists) flee*
Where ar they now? Summe are in Irelonde, — *Some*
In Walys other are, in myghti place, — *Others are in Wales*
And other han Caleys with hem to stonde: — *have Calais standing with them*
Thei robbeth and they reveth see and londe. — *plunder sea and land*
The kyng, or his ligeaunce or amytee, — *loyal people or friends*
Thei robbe anende, and sle withoute pitee. — *constantly; slay*

The Goldon Eagle and his Briddys III, — *three Birds (see note)*
Her bellys ha they broke, and jessys lorne. — *Their bells have; fetters lost*
The Silver Bere his lynkys al to fle, — *Bear flees his chains*
And bare is he behinde and eke beforne. — *also before*
The Lily-whit Lyoun, alas! forsworne
Is his colour and myght; and yet detrude
Entende thei the lond, and it conclude.[2]

fol. 19v Of bestialité, lo! ye so rude! — *Lo! Like beasts you are so uncivilized*
The noblis alle attende on the Antilope; — *Antelope*
Yourself and youris, ye yourself exclude,

---

1 Before 978: *The third part is on waging war and waging peace, / In which honor accompanies anything good*

2 Lines 996–98: *The Lily-white Lion, alas! treasonous / Is his color (i.e., not innocent) and might; and yet to subjugate / The land they intend, and to end it*

And lose soule and lyif. Aftir your coope *guilt*
Axe humble grace, and sette yourself in hope, *Ask [for]*
For and ye wiste hou hard lyif is in helle, *if you knew*
No lenger wolde ye with the murthre melle. *consort with murderers*

Ye se at eye, it nedeth not you telle, *You see by eye; not [be to] you told*
Hou that the beestis and the foulys alle, *birds*
That gentil are, ar sworn your wrong to quelle; *noble; end*
Ypocrisie of oothis wil not walle *oaths; protect*
You fro the sword, but rather make it falle
On your avarous evel governaunce, *avaricious*
That may be called pride and arrogaunce.

This geve I theim to kepe in remembraunce:
Goode Antilop, that eny blood shal spille, *Antelope*
Is not thi wille. Exiled is vengeaunce
From al thi thought; hemself, alas, thei kille. *themselves*
O noble Pantere! Of thi breth the smylle, *Panther; smell of your breath*
Swete and pleasaunt to beest and briddis alle, *birds*
It oonly fleth the Dragon fild with galle. *Dragon alone flees it, filled with gall*

What helpeth it, lo, th'angelis wil falle
On him with al our werreours attonys; *warriors at once*
Thei muste nede his membris al to malle. *must need to crush all his members*
Of this matere I stynte until eftsonys, *cease; later*
And fast I hast to write as it to doone is, *hasten*
That myght in right uppon the wrong prevaile
In londe and see, by knyghthode and bataile.

[Size of an Army (Veg. 3.1)]

fol. 20r Lo, thus th'electioun with exercise *selection; training*
And ordynaunce, as for a legioun, *formations*
Exployed is, as writeth olde wise. *Presented*
What ha we next? Belligeratioun. *have; Warfare*
O Jesse flour! Jhesu, salvatioun
And savyour, commaunde that my penne
To thin honour go right heryn and renne. *move quickly*

An oste, of exercise, *exercitus* *army, from exercising*
Hath holde of olde his name; a legioun *its name [in Latin]*
As an electioun is named thus,
And a choors of cohortatioun. *cohort*
The princys of her mynystratioun *their*

Her namys have, and aftir her degré	
The chevetaynys undir named be.	*chieftains*
Exercitus, that is to seyn an ooste,	
Is legiounys, or a legioun.	
Tweyne is ynough, and IIII is with the moste,	*Two; four*
And oon suffiseth in sum regioun.	
Therof, with ayde and horsmen of renoun,	*auxiliaries and horsemen*
As needful is, groweth good governaunce	
In every londe, and parfit prosperaunce.	*perfect*
What is an ayde? It is stipendiaryis	*auxiliary; paid men (i.e., mercenaries)*
Or souldiours conduct of straunge londe,	*brought from other lands*
To such a numbir as it necessary is.	
Aftir the legioun thei for to stonde	
In ordynaunce, to make a myghti honde.	*formation; mighty hand*
Heryn who wil be parfit and not erre,	*Of this; perfect*
Tak maysterys of armys and of werre.	*Achieve mastery; war*
fol. 20v This was the wit of princys wel appreved,	*approved*
And ofte it hath be seid and is conclude,	*has been said*
That oostis over-grete be myscheved	*over-large armies are undone*
More of her owne excessif multitude	*by their own*
Then of her foon, that thenne wil delude	*by their foes, who*
Her ignoraunce, that cannot modifie	*[Them because of] their; control*
The suffisaunce, an ooste togeder and gye.	*multitude, [of] an army; guide [it]*
To gret an oost is hurt in mony cace.	*Too large; many cases*
First, slough it is in journeyinge and longe:	*sluggish; long (i.e., it stretches out)*
Forthi mysaventure it may difface,	*mishaps may disorganize it*
Passagis hard, and floodis hye amonge.[3]	
Expense eek of vitaile is over stronge,	*also of victuals; too costly*
And if thei turne bak and onys fle,	*flee at once*
They that escape, aferd ay aftir be.	*will be ever after afraid*
Therfore it was the gise amonge the wise,	*guidance*
That of the werres had experience,	*wars*
Oonly to take an oost as wil suffice,	
Of preved and acheved sapience,	*proven; wisdom*
In chivalerys that han done diligence	*knights who had*
In exercise of werre. A lerned ooste	*experienced (i.e., veteran) army*
Is sure, an unlerned is cost for-loste.	*a cost already lost*

[3] *Rough roads, and high rivers at times*

In light bataile, oon legioun with ayde,
That is, X M^{l} men o fote, and too *10,000 footmen*
Thousand on hors, sufficed as thei saide. *2,000 on horseback*
They with a lord no grete estat to goo, *lesser lord*
And with a gret estate as mony mo.
And for an infinit rebellioun *long-term*
Twey dukys and tweyn oostys went adoun. *Two dukes and two armies*

[Maintaining Health in the Army (Veg. 3.2)]

fol. 21r
Provisioun be mad for sanytee *sanitization*
In watre, place, and tyme and medycyne *water-supply; season*
And exercise. In place, ther hath be *where has been*
The pestilence, his place anoon resigne. *this place immediately abandon*
To weet marice and feeld to hard declyne. *Over-wet marshes; fields too*
To high, to lough, to light, to derk, to colde, *Too; low*
To hoot, is ille. Attemperaunce be holde. *hot; Temperaunce*

In snow and hail and frost and wintir shouris, *showers*
An ooste beyng, most nedes kacche colde.
For wyntir colde affrayeth somer flourys, *wears away summer flowers*
And mareys watir is unholsom holde. *marsh; held unhealthy*
Good drinke and holsom mete away wil folde *wholesome foods will drive away*
Infirmytee; and fer is he fro wele, *far; from well-being*
That with his foon and sekenesse shal dele. *with [both] foes and illness; deal*

Cotidian at honde ha medycyne: *Ever at hand have*
First for the prince, as needful is his helth *[for] his health is as vital*
To th'ooste as to the world the sonne shyne; *as the sun's shining is to the world*
His prosperaunce procureth every welth. *prosperity*
But let not exercise goon o stelthe; *go neglected*
Holde ever it. Ful seelde be thei seek *Very seldom are they sick*
That ever uppon exercise seeke.

In over-colde and hoot, kepe thee covert, *keep yourself sheltered*
And exercise in tymes temperate:
Footmen in high and lough, feeld and desert; *Footmen [train]; low*
An hors to lepe a dich, an hege, a yate. *hedge; gate*
Tranquillité with peax and no debate *Relaxation; peace*
Be sadly kept, exiled al envie; *rigorously*
Grace in this governaunce wil multiplie.

[FEEDING THE ARMY (VEG. 3.3)]

fol. 21v Ha purviaunce of forage and vitaile — *Have provisions; victuals*
For man and hors; for iron smyteth not — *smites*
So sore as honger doth, if foode faile. — *as hunger does*
The colde fyer of indigence is hoote, — *destitution burns*
And wood theron goth every man, God woot; — *insane; God knows*
For other wepen is ther remedie, — *remedy [from wounds]*
But on the dart of hongir is to deye.

Or have ynough, or make a litil werre, — *Either; make [more] by raiding*
And do the stuf in placys stronge and sure. — *hold the materials*
In more then ynough, me may not erre; — *[having] more than enough, men*
The moneyles by chevishaunce procure — — *borrowing [must] procure [it]*
As lauful is, I mene, nat usure. — *lawful; not [through] usury*
But tak aforn the day of payment: — *payback before*
It loseth not, that to the prince is lent.[4]

What man is hool in his possessioun, — *secure; possessions*
If he ha no defense of men of armys? — *has; armed men*
Beseged if me be, progressioun — *if I am, further acquisition*
That ther be noon, and noo vitail in arm is, — *none; victuals are in hand*
O woful wight, ful careful thin alarm is! — *woeful man, full of cares is your*
Honger within, and enmytee abowte, — *[If] hunger [is]; enmity*
A warse foo withinn is then withoute. — *worse*

And though thi foo withoute an honger be, — *if your foe*
He wil abide on honger thee to sle; — *wait; slay*
Forthi comynge a foo, vitaile thee, — *Therefore; supply yourself*
And leve hym noght, or lite, unworth a stre; — *not worth a straw*
Whete and forage and flesh, fissh of the see, — *meat*
Wyn, salt, and oyle, fewel and every thinge — *oil, fowl*
That helpeth man or beest to his lyvinge:

fol. 22r Tak al, thi foo comyng, and mak an oye
That every man to strengthes ha ther goodis,[5]
As thei of good and lyves wil ha joye,
And negligentys to compelle it good is. — *it is good to compel [those who are] negligent*
The feriage be take away fro flodis, — *ferries; from rivers*
The briggis on the ryverys to breke, — *bridges*
And passagis with falling tymbour steke. — *blocked*

---

4 *What is lent to the prince is never lost*

5 Lines 1139–40: *Take everything, [if] your foe [is] coming, and make a proclamation / That every man must take his supplies to strong places*

The yatis and the wallys to repare, *repair*
The gunnys and engynys and tormente, *tormenta (i.e., torsion engines)*
And forge newe, ynowe if that ther nare; *enough; near*
Ful late is it, if thi foo be presente, *Too*
And fere ingoth, if hardinesse absente. *fear arises*
Bewar of this, and every thing provide,
That fere fle, and good corage abide. *So that fear flees*

Golde it is good to kepe, and make stoor *a stockpile*
Of other thing, and spende in moderaunce. *moderation*
More and ynough to have, it is not soor,
And spare wel, whil ther is aboundaunce: *save*
To spare of litil thing may lite avaunce. *help little*
By pollys dele, and not by dignitee, *By individual rations; rank*
So was the rewle in sage antiquytee.

And best bewar, when that thin adversary *your*
Wil swere grete, ye by the Sacrament, *Makes great oaths, even*
And use that, ye and by Seint Mary,
And al that is undir the firmament.
Beleve nat his othe! His false entent
Is this: thi trewe entent for to begile.
The preef herof nys passed but a while. *proof of this is shown soon enough*

fol. 22v Wel ofter hath fals simulatioun *Far more often has*
Desceyved us, then opon werre. And where *open war*
Me swereth ofte, it is deceptioun. *Men*
Judas, away from us! Com thou no nere! *nearer*
Thou gretest, Goddis child as thaugh thou were, *You weep, as if; God's*
But into thee is entred Sathanas, *Satan has entered*
And thou thiself wilt hange, an hevy cas!

[Preventing Mutiny (Veg. 3.4)]

Sumtyme amonge an ooste ariseth roore. *a riot arises*
Of berth, of age, of contré, of corage *birth (i.e., social class)*
Dyvers thei are, and hoom thei longe sore, *Diverse; they sorely long for home*
And to bataile thei wil, or out of wage. *be discharged*
What salve may this bolnyng best aswage? *assuage this tumor*
Wherof ariseth it? Of ydilnesse. *idleness*
What may aswage it best? Good bisinesse. *busy-ness*

With drede in oost to fight thei are anoyed, *fear of fighting the army is afflicted*
And speke of fight, when theim wer lever fle, *they would rather flee*

And with the fode and wacch thei are acloyed. *rations; wearied*
"Where is this felde? Shal we no batail see?
Wil we goon hoom? What say ye, sers?" "Ye, ye!" *go home; sirs*
And with her hed to fighting are thei ripe *their head (i.e., commander); ready*
Al esily, but he the swellinge wipe. *tumor [must] remove*

A remedie is, when thei are asonder,
The graunt tribune, or els his lieutenaunt, *great tribune (officer)*
With discipline of armys holde hem undir
Severously, tech hem be moderaunte, *Severely, teach*
To God devout, and fait of werrys haunte, *feats of war (i.e., military drills) repeat*
The dart, baliste, and bowe, and cast of stoon,
And swymme and renne and leep, tech everychoon. *swim; run; leap*

fol. 23r Armure to bere, and barrys like a sworde, *Armor; bars [made] like swords*
To bere on with the foyn, and not to shere, *carry; point; cut*
And smyte thorgh a plank other a boorde, *thrust through; board*
And myghtily to shake and caste a spere, *brandish*
And loke grym, a M[l] men to fere, *1,000 men to frighten*
And course a myghti hors with spere and shelde, *ride*
And daily se ho is flour of the feelde. *see who is the flower (best) in the field*

To falle a grove or wode, and make a gate *fell; wood*
Thorgh it, and make a dike, and hewe adoun
A cragge, or thurl an hil, other rebate *tunnel a hill, or reduce*
A clyf, to make an even regioun, *flat area*
Or dowbil efte the dike abowte a toun, *double the height of*
To bere stoon, a boolewerk forto make, *carry stones, a bulwark*
Other sum other gret werk undirtake. *Or some*

The chivaler, be he legionary, *knight, whether he [is a] legionnaire*
As seide it is beforn, on hors or foote,
Or aydaunt, that is auxiliary,
On hors or foot — if that thei talk or mote *discuss*
Of werre, and reyse roore, up by the roote *war, and raise riot*
Hit shal be pulde with myghti exercise *pulled*
Of werreourys, governed in this wise.

Commende, and exercise, and holde hem inne, *them in*
For when thei ha the verrey craft to fight, *have the very skills*
Thei wil desire it, wel this for to wynne.
He dar go to, that hath both art and myght. *dares go to [it], who*
And if a tale is tolde that eny knyght
Is turbulent other sedicious, *or*
Examyne it the duke, proceding thus:

fol. 23v The envious man, voide his suggestioun,
And knowe the trowth of worthi and prudent
Personys, that withouten questioun
Wil say the soth, of feith and trewe entent. *truth*
And if the duke so fynde him turbulent, *disorderly*
Dissever him, and sende hym ellys where, *Separate*
Sum myghti feet to doon as thaugh it were:

To kepe a castel, make a providence, *gather provisions*
Or warde a place, and do this by th'advyce *guard*
Of counsel, and commende his sapience, *wisdom*
That he suppose hymself heryn so wise, *[In order] that*
That therof hath he this honour and price. *distinction*
So wittily do this, that he, rejecte, *[though] rejected*
Suppose that to honour he is electe. *elected*

For verreily, the hole multitude *truly, the whole assembly*
Of oon assent entendeth not rebelle,
But egged ar of theim that be to rude, *are egged on by those who*
And charge not of heven or of helle, *have no regard for*
With mony folk myght thei her synnys melle; *they might mix their sins*
Thei were at ease her synnys forto wynne,
Suppose thei, if mony be therinne.[6]

But use not the medycyne extreme
Save in thin utterest necessitee. *Except in your most utter*
That is, the crymynous to deth to deme *to condemn the guilty to death*
The principals: by hem that other be *ring-leaders: by them (their example)*
Aferd to roore. Yet better is to se *Afraid to riot; see*
An oost of exercise in temperaunce
Obeysaunt, then for feere of vengeaunce.

[MILITARY SIGNALS (VEG 3.5)]

fol. 24r The werriours ha myche thing to lerne, *many things to learn*
And grace is noon, to graunte negligence,
Wher mannys helth is taken to governe. *men's safety*
To lose that, it is a gret offense.
And sikerly, the best diligence *certainly*
Unto th'onour of victory t'ascende,
The seygnys is or tokenys t'attende. *Are the signals*

[6] Lines 1242–43: *They are unconcerned to be allotted [their individual shares of] the sin, / They think, if there are many of them involved*

For in bataile, when al is on a roore, *in uproar*
The kynge or princys precept, who may here *king or commander; hear*
In such a multitude? And evermore
Is thinge of weght in hond, and gret matere, *in progress*
And how to doon, right nedful is to lere; *do [it]; learn*
Therfore in every oste antiquitee
Hath ordeyned III signys forto be. *3 kinds of signals*

Vocal is oon, and that is mannys voys. *a man's voice*
Semy-vocal is trompe and clarioun *Semi-vocal*
And pipe or horn. The thridde macth no noys, *makes no noise*
And mute it hight or dombe, as is dragoun
Or th'egil or th'image or the penoun,
Baner, pensel, pleasaunce, or tufte or creste
Or lyvereys on shildir, arm, or breste.[7]

Signys vocal in wacch and in bataile *watch*
Be made, as wacch woordis: "Feith, hope, and grace," *watch-words (i.e., camp passwords)*
Or "Help us God," or "Shipman, mast, and saile,"
Or other such, aftir the tyme and place. *according to*
Noo ryme or geeste in hem be, ner oon trace, *No rhyme or story; not one*
Ne go thei not amonge us, lest espyes *Nor should they go [unchanged]; spies*
With wepon of our owne out putte our eyis. *eyes*

fol. 24v Semy-vocals, as trumpe and clarioun
And pipe or horn, an hornepipe thoo
It myghte be; the trumpe, of gretter soun,
Toward batail blewe up "Go to, go to!"
The clarions techeth the knyghtys do,
And signys, hornys move. And when thei fight, *And signals [or] horns move [them]*
Attonys up the soun goth al on hight. *At once; goes on high (i.e., all over)*

To wacch or worch or go to felde, a trumpe *work*
Hem meved out, and to retourne; and signys *Moved them*
Were moved, how to do, by hornys crompe, *curved*
First to remeve, and fixe ayeyn ther digne is. *appropriate*
Oonly the clarioun the knyghtis signe is;
Fight and retrayt and chace, or feer or neer, *retreat; either far or near*
The clarion his voys declareth cleer.

[7] Lines 1268–71: *And it is called mute or silent, like a dragon [banner] / Or the eagle or the [emperor's] image or the pennon, / Banner, battle-streamer, knightly emblem, or [helmet] tassel or crest / Or the liveries (heraldic badges) [displayed] on the shoulder, arm, or breast*

What so the duke commaundeth to be doon — *Whatever; do*
In werk or wacch or feeld, or frith or werre, — *either peace*
At voys of these it was fulfild anoon. — *At the sound; at once*
The signys mute, in aventure a sterre, — *visual signals; perhaps a star*
A portcolys, a sonne, it wil not erre, — *portcullis; sun*
In hors, in armature, and in array — *On*
They signifie, and make fresh and gay.

Al this in exercise and longe usage — *training and long usage*
Is to be knowe. And if a dust arise,
Theere is an oost, or sum maner outrage. — *trouble*
With fiyr a signe is mad in dyvers wise — *fire; made in many ways*
Or with a beem. Uche in his contré gise — *[hanging] beam. Each; country's way*
His signys hath, and daily is to lerne, — *(i.e., they must be changed daily)*
That aftir hem men gide hem and governe.

[Caution When Near the Enemy (Veg 3.6)]

fol. 25r Tho that of werre have had experience,
Afferme that ther is in journeyinge — *traveling*
Gretter peril then is in resistence — *Greater peril than; fighting*
Of fers batail. For in the counterynge — *A fierce battle*
Men armed are oonly for yeynstondinge — *defense*
And expugnatioun of hem present — *overcoming of those*
In fight; theron oonly ther bowe hath bent. — *their bow is only bent on this*

Their sword and hert al preste ereither fight; — *are both pressed into the fight*
In journeyinge ereither lesse attente is. — *are less ready*
Assault sodeyne a day other by nyght,
For unavised men ful turbulent is. — *unprepared*
Wherfore avised wel and diligent is
The duke to be purveyed for unwist, — *prepared; unknown [things]*
And redy is the forseyn to resiste. — *foreseen [things]*

A journal is in every regioun — *An itinerary*
First to be had, wheryn he thinketh fight, — *expects to*
Wheryn have he a pleyn descriptioun
Of every place, and passage a forsight, — *an intelligence*
The maner, wey, both turnyng and forthright, — *straight*
The dale and hil, the mountayn and the flood.
Purtreyed al to have is holdon good. — *Illustrated; held*

This journal is to shewe dukys wise — *to [be] shown [to]*
Of that province, or as nygh as may be, — *as near [to it] as possible*

The purtreyture and writing forto advise. — *illustrations*
And of the contrey men a serch secré — *natives; secret*
Himself he make, and lerne in veritee — *and [also] learn the truth*
Of hem, that on her lyf wil undirtake, — *From them; on [peril of] their life*
That thus it is, and under-warde hem make. — *his loyal subjects make them*

fol. 25v Tak gidis out of hem, beheste hem grete, — *guides; promise them great [things]*
As to be trewe, her lyif and grete rewarde, — *truthful; their lives*
And other if thei be, with deth hem threte, — *otherwise; death*
And sette a wayt secret on hem, frowarde
Whethour thei thinke be other towarde;
Thei, this seynge, wil wel condite and lede,
Of grete rewarde and deth for hope and drede.[8]

Tak wise and used men, and not to fewe; — *experienced*
Good is it not to sette on II or III
The doubte of al, though thei be perfit trewe; — *fate of all, even if they are perfectly*
The simpil man supposeth ofte he be
Weywiser then he is, and forthi he — *More wise; therefore*
Behesteth that he can not bringe aboute; — *Promises*
And such simpilnesse is forto doubte.

And good it is, that whidirward goth th'ooste, — *wherever the army goes*
Secret it be. The Mynotaurys mase — *Minotaur's maze*
Doctryned hem to sey: "Whidir thou gooste,
Kepe it secret. Whil thi foomen go gase — *gaze*
Aboute her bekenys, to tende her blase, — *their watchfires; their blaze*
Go thou the way that thei suppose leeste — *least*
Thou woldest go. For whi? It is sureste."

Espyis are, of hem bewar! Also — *[There] are spies*
The proditours that fle from oost to ooste, — *traitors*
Bewar of hem, for swere thei never so, — *even if they swear never [to do] so*
They wil betray, and make of it their booste. — *boast*
Escurynge is to have of every cooste; — *Reconnoitering; coast*
Men wittiest on wightest hors by nyght — *strongest*
May do it best, but se the hors be wight. — *strong*

fol. 26r In a maner himself betrayeth he,
Whos taken is by negligence th'espie.[9]
Forthi bewar, and quicly charge hem se — *Therefore; look out*

---

8 Lines 1338–41: *And set a secret watch on them, [such that it is] behind / When they think it is in front; / They, seeing this, will well guide and lead [you], / Either in hope of great reward or in fear of death*

9 Lines 1363–64: *In a sense he betrays himself, / He who is ambushed due to neglecting scouts*

On every side, and fast ayeyn hem hye. — *return*
Horsmen beforn eke ever have an eye; — *in front*
On uch an half footmen, and cariage — *On both flanks footmen; baggage*
Amyddis is to kepe in the viage. — *journey*

Footmen it is to have and of the beste
Horsmen behinde; uppon the tail a foo — *upon the rear a foe*
Wil sette among, and sumtyme on the breste, — *front*
And on the sidis wil he sette also.
With promptitude it is to putte him fro. — *speed he must be repulsed*
Light herneysed, and myghtiest that ride,
Doubte if ther is, putte hem uppon that side.

And archery withal is good to take,
And if the foo falle on, on every side,
Good wacch on every side it is to make.
Charge every man in herneys fast abide, — *Order; to stay in secure harness*
And wepynys in hondys to provide. — *weapons*
Selde hurteth it, that is wel seyn beforn, — *Seldom does it hurt anything*
And whos is taken sleping, hath a scorn!

Antiquitee provided eek, that roore — *chaos*
Arise not in th'oost, for trowbelinge — *the army*
The chivalers behinde other before, — *knights*
As when the folk that cariage bringe, — *lead the baggage*
Ar hurt, or are aferd of on comynge, — *afraid a [foe] coming*
And make noyse. Herfore helmettis wight — *Therefore strong helmets*
A fewe uppon the cariours were dight. — *carters were given*

fol. 26v A baner hadde thei togedre to,
Alway CC undir oon banere; — *200 under each banner*
The forfighters asondred so therfro, — *fighters in front were separated from them*
That no turbatioun amonge hem were, — *confusion*
If that ther felle a conflicte enywhere.
And as the journeyinge hadde variaunce,
So the defense had divers ordynaunce: — *diverse*

In open felde, horsmen wold rather falle — *fields, horsemen*
On then footmen; in hil, mareys, and woodis, — *marshes*
Footmen rather. In feeld and frith to walle — *field and woods; protect*
An oost with myght, as wil the place, it good is, — *as well as*
And to bewar that slough viage or floodis. — *muddy road or rivers*

Asondre not the chivalerys, for thynne
If that me be, ther wil the foo bygynne.[10]

Therfore amonge it is to sette wyse
Doctours, as of the feelde, or other grete:
The forgoer to sette unto his sise,
And hem that beth to slough, forthward to gete.[11]
To fer aforn, and sole, a foo may bete; *Too far ahead; alone; beat*
He may be clipped off, that goth behinde. *cut off, who goes behind*
And to goon hole as o man, that is kynde. *So to go together; proper*

In placys as him semeth necessary,
An adversaunt wil sette his busshement, *ambush*
Not in apert, but in covert to tary, *open space; in secrecy; wait*
And falle uppon. The duke heer diligent
It is to be, to have his foomen shent; *foes harmed*
But every place it is the duke to knowe,
So that his witte her wylis overthrowe. *their plots*

fol. 27r If thei dispose in mountayn oponly *they array*
T'assaulte, anoon ha prevely men sent *To assault, at once have secretly*
To an herre hil, that be therto neer by, *higher; nearby*
And so sette on, that of the busshement
Above her hed, and of thiself present. *their heads*
Thei be aferd, and sech away to fle, *[will] be afraid; seek*
When over-hede and in the frount thei se. *they see [you]*

And if the way be streyt and therwith sure, *road; safe*
Let hewe adoun aboute, and make it large: *hew down [growth] around it*
In large way, peril is noo good ure. *peril is uncommon*
Also this is t'attende as thinge of charge *to be practiced as an important matter*
(Ye, rather then governe ship or barge):
That wher the foo by nyght other by day *whether*
Is used oon to falle and make affray, *typically*

And voyde that. It is to seen also,[12]
What is his use, on hors outher o foote, *preference; or on foot*
With fele or fewe his feetys for to doo, *many or few his feats*
That sapience his werkys alle unroote. *wisdom (i.e., forethought); render useless*
Of balys also grete is this the boote: *even great evils; remedy*

[10] Lines 1403–04: *Do not separate the cavalry, since wherever the [lines of] men are thin, the enemy will begin the attack there*

[11] Lines 1407–08: *The officer in front to set his pace, / And those who are too slow, to bring them forward*

[12] *And avoid that (i.e., traveling at that time). It is also to [be] seen*

Dayly to gynne go in such an hour — *begin to march*
As may be sure both oost and governour.

And yet bewar of simulatioun. — *deception*
To festeying calle in sum fugitif — *feasting; deserter*
And here him wel with comendatioun,
And lerne first, hou fellen thei in strif, — *how they fell in trouble*
And him beheste an honorabil lif. — *promise him*
Lerne of him al, and thenne aday or nyght, — *everything [about the enemy]*
When thei suppose leest, mak hem afright.

[How to Cross a River (Veg 3.7)]

fol. 27v
Agreved ofte are oostis negligent,
When it is hard passage over the floodys, — *rivers*
For if the cours be over violent
Or over deep, gret peril in that flood is.
A remedy to fynde heryn right good is,
For hevy men, pagis, and cariage — *pages*
Ar drowned oftyn tyme in such a rage.

The depth assay, and make of horsys hye — *Test the depth; go*
Tweyne eggys: oon be sette ayenst the streem, — *Two lines; current*
The myght therof to breke; another plye — *strength of it to break; set*
Benethe that, t'awayte uppon the fleem — *flow*
And charge theim, that thei attende on hem — *order them; watch for those*
That faile foote, and brynge theim alonde, — *Whose footing fails*
And thus til th'ooste be over, shal they stonde.

The flood is over deep in playn cuntré, — *[If]; flat country*
Departe it ofte, and make it transmeabil: — *fordable (i.e., more shallow)*
That most be doon with dykis gret plenté. — *many ditches*
And wil it not be so, sette ore a gabil,
On empti vesselling ley mony a tabil[13]
Fro lond to lond a brigge is made anoon,
And sure ynough it is for hors and mon. — *secure; horse and man*

Horsmen have had of reed or seggis shevys, — *bundles made of reeds or sedges*
Theron carying their armure as thei swymme,
But better is, to voiden al myschevys, — *avoid all accidents*
Ha skafys smale, and hem togedir trymme — *small skiffs*
With coorde alonge, atteynynge either brymme, — *attached to either side*

[13] Lines 1464–65: *And if that will not work, set over [it] a thick rope: / [And] on empty barrels lay many planks*

	And anchore it and tabil it at large,	*cover*
	And sure it is as arch or shippe or barge.	*secure; bridge*
fol. 28r	Yet war the foo, for uppon this passage	*beware*
	He leyt awayt! Anoon thin ooste dyvide	*lies in wait! At once*
	And stakys picch, encounter their viage,	*drive [into the ground]*
	And in that stede, if good is thought t'abide,	*place; to wait*
	Mak up a strong bastel on eyther side,	*bastion*
	And there, as axeth chaunce, it is to stonde	
	And ha vitaile out of ereither londe.	*have victuals from either bank*

[HOW TO BUILD CAMP (VEG. 3.8)]

	Now castellinge in journey is to write.	*fortifying an encampment*
	Not everywhere is founden a citee,	
	An ooste to loge, and vilagis to lite	*lodge; villages too little*
	For it ther ar, and siker thei ne be,	*secure they are not*
	As, to be sure, it is necessitee	
	To take a grounde as good as may be fonde,	
	And theruppon to make our castel stonde.	
	Leve not the better grounde unto thi foo.	
	Bewar of that se, watir, ayer, and londe	*Be sure*
	Holsom be there, and foode ynough ther to	*Wholesome*
	For man and hors, and woode ynough at honde.	
	No force if rounde or anguler it stonde,	*It matters not*
	But feyrest is the place and moost of strengthe,	*Though the best*
	When twey in brede is thryis in the lengthe.	*two in breadth; thrice (i.e., 2x3)*
	Mesure a grounde, as wil thin ooste suffice.	
	To wide it is, thin ooste therin is rare.	*[If] it is too wide, your; sparse*
	To streyt, thei be to thicke. A myddil sise	*Too narrow; size*
	Is beste. Now make it up, no labour spare.	*establish it*
	It mot be doon, theryn is our welfare!	*It must be done*
	As for a nyght, mak up of turf a wale	*a turf wall*
	And stake it, on our foo the poyntis t'avale.	*against our foe directing the points*
fol. 28v	A turf it is, when gras and herbe is grave	*pulled*
	Up with the grounde, with irons mad therfore.	*made [by] iron tools*
	A foote brode, a foote and half it have	
	In lengthe, and half a fote thick, no more.	*foot*
	But if the lond solute be, not herfore	*earth is loose, not of that kind*
	Turf like a brik to make of necessary,	
	Thenne is to make a dike tumultuary.	*temporary (hastily raised) ditch*

Make it III foote deep, and V obrede, *3 feet deep and 5 [feet] wide*
And stake it as beforn, utward to stonde; *angled outward*
O nyght to dwelle heryn it is no drede. *One*
And if thi foo be nygh, him to yeynstonde, *near; withstand*
A gretter werk it is to take on honde.
Sette up in ordir every man his sheeld,
Whil princys and prudentys parte a feeld. *officers and instructors mark out*

Uch centyner take up the werk footmel, *century; a footage of the work*
With sword igord, anoon caste up the dich, *armed; ditch*
And IX foote obrede wil do wel, *9 feet wide*
XI is as good; but poore and rich *11 feet; poor and rich [alike]*
Most on this werk, and even worch ilich, *Must; work alike*
XIII foote obrede or XVII *13 or 17 feet wide*
Is best of alle a werre to sustene.

The numbir odde is ever to observe,
And hege it, other stake it up to stonde, *raise it, or*
Therto ramayle and bowys ar to kerve, *small branches; boughs; carve*
Areyse it to his hegth above londe,
And make it castel-like with myghti honde,
With loupis, archeturis, and with tourys. *loopholes, barbicans; towers*
O chivalers! In this werk your honour is

.fol. 29r X footemel the centeneris take *A length of 10 feet*
This werk to doon, and ther uppon attende,
That every company his cant up make *share*
And stynte not, until a perfit ende *cease; perfect*
Of al be mad. And who doth mys is shende. *wrong; punished*
Forwhi? The prince himself goth al aboute
And by and by behaldeth every rowte. *company*

But lest assault felle on hem labouringe, *fall on those*
The hors, and thei on foote of dignitee,
That shal not worch, in circuyte a rynge
Shal make, and kepe off al hostilitie. *prevent all*
And first, as for the signys, majestie
Assigne place, for more venerabil
Then thei, ther is nothing, this is notabil.[14]

And aftir that, the duke and erlys have *earls*
The pretory, a grounde out set therfore, *praetorium (i.e., headquarters)*

[14] Lines 1542–44: *And first, as for the standards, the highest [leaders are] assigned their headquarters, for it is known that nothing is more venerated than they are*

And for trybunys out a grounde thei grave, — *tribunes*
Her tabernaclis thei theryn t'enstore — *tents; pitch*
For legions and aydis, lesse and more, — *auxiliaries*
On hors other o foote; a regioun
And place is had to picch her paviloun. — *their*

And IIII on hors and IIII o foote a-nyght — *4 on horseback and 4 on foot each night*
In every centeyn hadde wacch to kepe, — *century; watch*
And it deperted was, to make it light, — *divided*
That reasonabil tymys myght thei slepe.
For right as houris aftir houris crepe,
So went the wach, and kept his cours aboute,
Footmen withinne, and horsed men withoute.

fol. 29v Thei go to wacch by warnyng of the trumpe, — *trumpet*
And there abide until their houris ende.
Away thei go, by voys of hornys crumpe. — *curved horns*
A wacch of serch also ther was t'attende — *A check on sentries*
That wel the tyme of wacchinge were spende.
Trybunys made of theim th'electioun,
That hadde of al the wacch directioun. — *command*

And twye a day the contrey was escured — *twice a day; scoured*
By horsmen, in the morn and aftirnoon;
Not by the same alway, for that endured
Shuld not ha been. This feleship hath doon: — *have*
They most reste, and other wynne her shoon. — *win their shoes (i.e., prove themselves)*
Thus bothe man and hors may be releved,
Ye, ofte ynough, and not but litil greved.

And on the duk hangeth the governaunce,
That in this castellinge he ha vitaile — *encampment he has victuals*
For every wight withoutyn variaunce, — *man*
Clooth, wepon, herneysing, that nothing faile. — *harnessing*
And in fortressis nygh it is availe — *[smaller] forts nearby*
Footmen to have and hors; ferde is thi foo, — *afraid is your foe*
If thou on every side uppon him goo.

[Choosing Battle or a Raid (Veg. 3.9)]

Mortal bataile in hourys II or III — *2 or 3 hours*
Termyned is, and hope on that oon side — *Is ended*
Is al agoon. But a good prince is he
That can him and his ooste so wisely gide,

With litil slaught to putte his foo fro pride, — *little slaughter*
Pluck him unwar, and fray his folk to renne — *Catch him unaware; frighten; run*
Away, and myghtily sette aftir thenne.

fol. 30r
On this behalve it is ful necessary,
That olde and exercised sapience — *trained wisdom*
The duke to counsel have, and with hem tary, — *stay*
As wil the tyme, and here their sentence — *hear; advice*
Of vinqueshinge covertly by prudence — *vanquishing craftily*
Or by apert conflict, that is, bataile; — *open*
The surer way to take and moost availe.

Here hem heryn, and what folk hath thi foo, — *Hear them*
And charge that thei glose not, for it — *flatter*
Doth oftyn harm. And here theim also — *hear them*
Speke of her exercise, her strength and wit, — *their*
And to their adversayrys how thei quyt — *acquitted*
Hemself aforn, and whether his horsmen — *Themselves before*
Be myghtier in fight, or his footmen.

Also the place of conflicte is to lerne,
And what thi foo himself is, what his frendis;
Wher he be wys a werre to governe, — *Whether*
And whar thei lyve as angelis or fendis, — *whether; fiends*
Wher variaunt, or uchon others frend is,[15]
And wher thei use fight in ordynaunce — *typically fight in formation*
Or foliously, withoute governaunce. — *incautiously*

And every poynt forseyd, and other moo, — *aforesaid; more*
Considir in thin oost, and tak avis — *advice*
Of hem, what is the beste to be do.
And peyse al in balaunce, and ay be wys. — *weigh everything; always be wise*
And if thin ooste is ace, and his is syis, — *an ace; a six*
What so thei sey, covertly by prudence
Dispose thee to make resistence.

fol. 30v
Dischere nat thi folk in eny wise: — *Despair not*
The ferde anoon is redy for to fle. — *fearful are soon ready*
Be vigilaunt and holde inne exercise,
And se thin hour. Ful oftyn tyme hath he — *await your time. Very often*
The herre hand, that kepeth him secré. — *higher hand (i.e., victory)*
Avaunte not for colde ner for hete, — *Boast; nor*
For smale dooth that speketh over grete.

---

[15] *Whether at variance or on good terms with one another*

[HOW TO LEAD AN INEXPERIENCED ARMY (VEG. 3.10)]

Certeyn it is, that knyghthode and bataile
So stronge is it, that therby libertee
Receyved is with encreste and availe: *growth and aid*
Therby the croune is hol in majestee *crown is held*
And uche persone in his dignitee.
Chastised is therby rebellioun,
Rewarded and defensed is renoun.

Forthi the duke, that hath the governaunce,
Therof may thinke he is a potestate, *great dignitary*
To whom betakyn is the prosperaunce *prosperity*
Of al a lond and everych estate.
The chivalers, if I be fortunate,
The citesens, and alle men shal be *citizens*
If I governe wel, in libertee.

And if a faut is founden in my dede, *fault; deeds*
Not oonly me, but al the commyn wele *commonwealth*
So hurteth it, that gretly is to drede
Dampnatioun, though no man with me dele. *Condemnation*
And forthi, negligence I wil repele *disavow*
And do my cure in feithful diligence *best*
With favoraunce of Goddis excellence.

fol. 31r If al is out of use and exercise, *inexperienced and untrained*
As forto fight in every legioun
Chese out the myghtiest, the wight and wise,
And aydis with, of like condicioun. *[among] auxiliaries also*
With their avice unto correctioun *advice*
Reduce it al by his auctorité
The duke, and use a grete severitee. *severity [in punishment]*

Amended al as sone as semeth thee, *as soon as you can*
Make out of hem a stronge electioun:
Disparpiled lerne if thi foomen bee,[16]
And when thei lest suppose in their reasoun, *least think it possible*
Fal on, and putte hem to confusioun. *Fall upon [them]*
Therof thi folk shal take an hardinesse
And daily be desirous on prowesse.

[16] *If you learn that your foes are disorganized (i.e., in foraging)*

At brigge or hard passage or hillis browe *[a] bridge; hill's pass*
Is good to falle uppon. Or if ther be
Mire or mareys, or woode or grovis rowe *marsh; a row of trees*
Or aggravaunt other difficultee,
To falle uppon is thenne utilitee
The hors to sech. Unarmed ar aslepe; *horsemen; seek*
To falle uppon is good to take kepe. *remember*

Thus hardy hem, for whos is unexpert *This emboldens them*
Of werre, and woundis seeth, and summe slayn, *sees wounds, and some slain*
He weneth every strok go to his hert, *feels*
And wiste he how, he wolde fle ful fayn. *If he knew how; flee quickly*
But and he fle, retourne him fast agayn. *if he flees*
Thus with severitee and good usage
Ther wil revive in theim a fyne corage. *worthy courage*

fol. 31v Dissensioun among foomen to meve, *foes to stir*
Be thei rebellious or myscreaunt,
It is to do; theimselven thei myscheve. *they may harm themselves*
The traditour Judas was desperaunt; *traitor; despairing*
Himself he hynge. So wulle thei that haunt *hanged; will they; engage in*
Rebellioun or ellis heresie.
Alas! to fele thus wil lyve and deye. *too many; die*

Oon thinge heryn is wisely to be seyn, *herein (i.e., in this book); seen*
Of this matier that ther no man dispayre: *despair*
As hath be doon, it may be doon ayeyn;
A desolat castel man may repayre.
In wynter colde, in somer dayis fayre
Is good to se. So fareth exercise *(see note); fares [the] practice*
Of knyghthode and of werre, as seyn the wise.

In Engelond til now was ther no werre *there were no wars*
This LX yere, savynge at Seynt Albane, *For 60 years, except; St. Albans*
And oon bataile aftir the blasing sterre, *comet*
And longe on hem that whirleth as the fane.[17]
Is not their owne cryme her owne bane? *destruction*
Ther leve I that, and sey that exercise *readiness*
Of werre may in peax revyve and rise. *restore and raise up peace*

Seyde ofte it is: the wepon bodeth peax, *weapon presages peace*
And in the londe is mony a chivalere,
That ha grete exercise doubtlesse

[17] *And to them belongs the fault who whirl like the weathervane (i.e., the Yorkists)*

And think I wil that daily wil thei lere, *learn*
And of antiquitee the bokys here, *hear*
And that thei here, putte it in devoyre, *what they hear; into practice*
That despetaunce shal fle comynge espoyre. *despair; hope*

fol. 32r More esily a thing is al mad newe
In many cas, then is an olde repared; *than*
The plauntys growe, as olde tren up grewe, *trees*
And otherwhile a riche thing is spared. *sometimes somthing splendid*
It nedeth not to crave this declared,
But go we se, what helpeth to prevaile
Uppon the feelde in sette apert bataile.

[ON THE DAY OF BATTLE (VEG. 3.11)]

Here is the day of conflict uncerteyn,
Here is to se deth, lif, honour, and shame.
Glade us, o Lord, this day and make us fayn, *Gladden; rejoice*
And make us of this grete ernest a game! *seriousness*
Lord, make in us magnificent Thi name,
Thin angelis commaunde in us t'attende, *to wait upon us*
And she, Thi Modir, have us recommende.

Now is the duke the rather diligent, *more especially*
That forth he goth bytwene espoyre and drede. *hope and dread*
Now glorious the prince is sapient; *[is] the prince [who] is wise*
Now th'ignoraunt shal deye or harde spede. *fare badly*
In this moment manhode and knyghtly dede *deeds*
With Goddis honde is oonly to prevaile. *God's hand*
Now let se first, how wil our foon assaile. *foe make the assault*

The chivalers set forth first at the yate, *gate*
Whether ye dwelle in castell or citee,
And sette a frount or eny foo come ate, *before*
Til th'ooste come out undir securitee.
Go not to fer ne faste, for ye se, *too far nor too fast*
A wery wyght hath spended half his myght, *weary man*
And with the fresh is hard for him to fight.

fol. 32v And if thi foo the yatis ha forsette, *foe has already beset the gates*
Delay it and attende what thei mene. *observe what they intend*
Let hem revile and gnaste and gomys whette,[18]

[18] *Let them hurl insults and gnash their teeth and beat their gums (i.e., talk ineffectively)*

And breke her ordynaunce, and when thei wene — *break their formation; think*
Ye be aslepe, and they foryeton clene,
Breke on hem unavised day or nyght: — *by surprise*
This wisdom is to do, manhode and myght.

[Judging the Temper of the Men (Veg. 3.12)]

It is to frayne also with diligence, — *[best] to inquire*
Wher chivalerys think it be to fight,
Her countynaunce of fere or confidence — *Their measure of fear*
Wil be the juge: and truste not the knyght
That is aferd, ner hym that of his myght — *nor*
Presumeth. Inexpert what is bataile, — *Boasts. Those inexperienced in*
Conforte hem yet. Telle hem thei shal prevaile,

And reasounynge reherce rebellioun
Or myscreaunce, and how thei be forsake — *[will] be stripped*
Of alle goode. A prynce as a lyoun
May telle that aforn thei ha be shake; — *speak before those [who] had been shaking*
And if he may with reasounynge awake
An hardinesse in hem he may procede — *A boldness*
And ellys uttirly he stont in drede. — *otherwise he stands in danger*

The first sight is ferdfullest for tho — *most fearful for those*
That never were in fight; and remedie
Is in beholdinge ofte uppon her foo
Out of a siker place or placys heye. — *secure; high places*
Confort therof comyng, dispayr wil deye, — *despair will die*
Eke issuynge on hem with a prevaile — *issuing against them; advantage*
Is hardyinge to falle to bataile. — *Emboldens [them]*

[Selecting the Battleground (Veg. 3.13)]

fol. 33r Part of the victory is for to chese — *choose*
The herre grounde, and ay the herre it be, — *higher; always*
The more myght thou hast thi foo to ceese, — *you have to stop your foe*
And more sharp dounward the taclys fle, — *arrows fly*
Thi foon her fight is with the grounde and thee; — *The fight of your foes; you*
Yet footmen hors, and hors footmen t'assaile,
Theire is the cleef, the playn is hem t'availe.[19]

[19] Lines 1760–61: *Furthermore, [if your] footmen are to attack horsemen, or [your] horsemen to attack footmen, / The former is helped by cliffs (i.e., rough ground), the latter is helped by plains*

[Ordering the Battle Lines (Veg. 3.14)]

And if thou may ha with the sonne and wynde, *have with you sun*
Ereither on the bak is grete availe, *Either at the back*
Ereither also wil thi foomen blynde. *Or*
Ayeinst the wynde to fight, it is travaile, *Against*
A cloude of dust wil therwithal assaile
Thi foomen in the frount, and stony hem so *paralyze them*
That they her wit shal seke what to do. *shall search their wit to know*

Forthi the prince it is be provident *ought to have fore-thought*
And have a sight to wynde and dust and sonne,
And on the turnyng take avisement, *positioning*
Remembering hou certeyn hourys ronne: *hours run (i.e., time changes things)*
It wil not stonde, as stood when thei begonne. *remain, as it stood; began*
West wil the sonne and happely the wynde,[20]
But seen he wil that thei come ay behinde, *he (i.e. the prince) will see*

And ever smyte his foomen in the face.
And there an ende of that. Now wil we se,
This ooste embateled uch in his place, *engaged each in his position*
That noon errour in eny parti be:
Therof wel ordeyned utilitee. *well-ordered*
Wil nede arise, and his inordynaunce *needs*
May brynge (as God defende!) us to myschaunce.

fol. 33v First is to sette a frounte, an ege his name *[battle-]front, a line*
Is. Whi? The foon it shal behalde and bite, *It shall behold and strike the enemy*
Ther chivalers, the worthiest of fame, *There knights*
That wil with wisdom and with wepon smyte, *direct [themselves]*
Noo knyght apostata, noon ypocrite: *apostate; hypocrite*
Feers, feithful, ofte appreved, olde, and wise *Fierce; veteran*
Knyghtys be thei, none other in no wise. *no way*

This ege in dayis olde a principaunt: *line; a first-line (principes)*
Of wurthi men, as princys, had his name.
In th'ordre next personys valiaunt,
Such as ha sought honour and voyded shame *have; avoided*
That ure have had, to make her foomen tame, *typically; their foes*
Sette hem theryn, armure and shot and spere
That myghtily can use and wel bewere. *defend*

[20] *The sun will [move] west, and the wind [will move] by chance*

Next to the firste frount this is secounde,
    And as of old thei called hem hastate *hastati*
Bycause of use of spere and shaftis rounde.
    Of armure is noon of hem desolate.
III foote atwene had every man his state, *3 feet between*
    So in a M^l pace o length stood fixe *a length of 1,000 paces*
    A M^l DC LX and VI. *1,666 [men]*

Footmen were alle these, and stode in kynde
    In duble raunge, and everych hadde III *two lines; each had 3*
Foote, as byforn is seide, and VI behinde *6 behind (i.e., between the lines)*
    The raungis hadde a sondir, so that he *lines were offset*
That stood beforn, unlatted shulde be *unfettered*
    To drawe and welde his wepon, and to take
    His veer to lepe or renne, assaut to make. *leap or run*

fol. 34r In tho tweyn orderys wer ripe and olde *two lines*
    Appreved werryours of confidence,
That worthi men of armys had ben holde,
    With wighti herneysing for to defense. *harnessing*
These as a wal to make resistence
    Ay stille stode, hem may noo man constreyne *Always*
    T'avaunce forth or reere o foote ayeyne. *To move forward or back one*

Thei trouble not, lest other troubled were,
    But fixe abide, and welcom th'adversary *wait motionless*
With sword and axe, with shot and cast of spere,
    Until thei yeve her coors to seyntewary,[21]
Or fle. For whi? Thei dar no lenger tary.
    Thenne aftir hem that ar to go for al, *them (i.e., those in flight)*
    For these stille abide as doth a wal.

Tho tweyne eggys ar clept 'the grete armure,' *two lines are called*
    And aftir hem the thridde cours is sette
Of wighte and yonge and light herneysed sure,
    With dartys and with taclis sharpply whette, *arrows; whetted (i.e., sharpened)*
In dayis olde thei ferentayris hette. *were called ferentarii*
    The firthe cours was called the scutate, *fourth; scutati*
    Spedy to renne and glad to go therate. *run; go out*

Wight archery with hem to shote stronge,
    The yongest and the best and lustyeste
Archers with crankelons and bowys longe. *crossbows and long bows*

[21] *Until they give their (i.e., the enemy's) bodies to sanctuary (i.e., burial ground)*

	The ferenters and thei togedir keste	*put together [are]*
	Named the light armure, as for the beste	
	Thorgh shulde passe and first with shot provoke	
	The adverse part, and on hem reyse a smoke.	*raise a smoke (i.e., cause trouble)*
fol. 34v	If foomen fle, thei and horsmen the chase	
	Go swift uppon, and ellis thei retrete	*otherwise*
	And thorgh the frount indresse hem to their place.	*through the front [line] return*
	The grete armure, if thei com on an hete,	*come in an attack*
	Is hem to yeve of sword and axis grete.	*give*
	On hem the feeld is now for to defende.	
	Thei gynne wel, God graunte hem a good ende!	*begin*
	The fifthe cours was the carrobaliste,	*carroballistae*
	Manubalistys, and fundibulary	*Manuballistae; fundibulatores*
	And funditours, but now it is unwiste,	*fustibalii; unknown*
	Al this aray, and bumbardys thei cary,	*bombards*
	And gunne and serpentyn that wil not vary,	*[a] gun and serpentine*
	Fouler, covey, crappaude, and colveryne,[22]	
	And other soortis moo then VIII or IXne.	*more than 8 or 9*
	Heer faughte thei, that hadde as yet no sheelde,	*Here*
	As bachelers, with shot of dart or spere.	*recruits*
	The sixte cours, and last of al the feelde	
	Wer sheeldys, of the myghtiest that were,	
	The bellatourys beste in every gere;	*warriors*
	Antiquytee denamed hem triayrys,	*triarii*
	In theym, as in the thridde, al to repayre is.	*(see note)*
	Thei to be sadde in strength and requyete,	*resolute*
	More fervently to make invasioun,	
	To take her ease in ordir alwey seete,	*(i.e., be well-rested)*
	And if aforn wer desolatioun,	*if [those] ahead were destroyed*
	In theym therof was reperatioun:	*recovery*
	In eny part if ther wer desperaunce,	*desperation*
	Thei turned it anoon to prosperaunce.	*quickly to success*

[Spacing the Lines of Battle (Veg. 3.15)]

fol. 35r	Now the podisme — as whos wil sey, the space	*podismus*
	Of grounde upon to fight — it to se:	
	Aforn is seide, hou in a M[l] pace	*Before; 1,000 paces*

22 *Veuglaire, covey, crapaudeau, and culverin*

XVI C LX and VI may be, — *1,666 [men]*
So chivalers everych ha footis III — *each have 3 feet*
To stonde upon a foote and VI abacke — *6 [feet] behind them*
That for his veer and leep no rowme hym lacke. — *dodge and leap no room*

VI eggys heer sette in a M^{l} pace — *6 lines set here*
Shal holde II and XL^{ti} feet in brede, — *42 feet in breadth*
And so X M^{l} wil this grounde embrace. — *10,000 [men]*
Thus t'embataile is sure, and fer fro drede! — *to fight; far from fear*
And to II M^{l} pas III cours for nede — *3 lines would need 2,000 paces*
In long goth out, so that the latitute — *depth*
In XXI foote itself enclude. — *21 feet*

As here is taught, X M^{l} men may stonde — *10,000*
In oon or ellys in II M^{l} pace, — *1,000 or 2,000 paces*
And XX^{ti} M^{l} in the double londe, — *20,000*
And XXX^{ti} M^{l} in the threfolde space, — *30,000*
And XL M^{l} IIII folde is t'embrace; — *40,000 fourfold*
And this mesure is named the podisme, — *podismus*
Untaught in *Doctrinal* or in *Grecisme*.

A prince heryn expert, and hath to fight — *herein*
His feelde and of his folk the multitude,
Shal seen anoon how thei shal stonde aright, — *stand rightly (i.e., be organized)*
And if the feeld is short and brod, conclude — *broad*
On rangis IX, and by this similitude, — *nine lines*
Be short and huge in brede, or longe and rare. — *breadth; thin*
But myghtier is brede, and mo may spare.

fol. 35v And rare, an ooste if th'adversary seeth,
He breketh on with hurt peraventure,
Wher thicke outholdeth him ayenst his teeth.[23]
And ther an ende of that. But hoo shal cure — *whoever shall care*
Ereither, horn and myddis, to be sure, — *horn (i.e., wings) and center*
Ordeyne that, or aftir dignitee — *either*
Or aftir th'adversayris qualitee.

[DEPLOYING HORSEMEN (VEG. 3.16)]

The feelde o foote ordeyned in this gise, — *of footmen arranged; manner*
To sette it is these hors at eyther horn, — *horsemen [placed] at either wing*
As writeth in her werkys olde wise, — *their works (i.e., books)*

[23] *Whereas thickness holds out against his teeth (i.e., aggression)*

That herneysed sperys be sette aforn, *harnessed spearmen; in front*
Unharneysed abak, that of be born *in order to carry*
The storm fro theym, whil myghti hors defende *fight from them*
Stronge archerye o foote to shote on ende.

For to defende have horsis myghtieste,
Tho hornys in attempting is to sende *wings*
Out hors the swiftest and the wightieste, *most powerful*
To trouble theym sette on a pace on ende.
The duke it is to knowe and comprehende,
What hors ayenst what throngys ar to goon, *against which throngs*
And whar he have hors as goode as his foon. *foes*

Their hors ar over us, theryn is boote: *[If] their horsemen outnumbers; remedy*
Tak wight and yonge men with sheeldis light,
With twene on hors, sette one of theim o foote.[24]
With hem resiste our adversayrys myght.
But this to take effecte and spede aright, *for this; go well*
These yonge men herof grete exercise *training*
Moste have, as telleth werreourys wise. *Must; warriors*

[RESERVES (VEG. 3.17)]

fol. 36r And aftir al his ooste, a duke shal have *behind all his army*
A myghti choyce of men on hors and foote,
Ereither horn and breste for to save, *Either wing or front (i.e., center)*
That if the boorys hed in wolde wrote, *boar's head roots around in the trees*
A sharre shere his groyn of by the roote. *Shears [will] shear his snout off*
The boorys hed is a triangulere
Of men, a boorys hed as thaugh it were.

If that come on, with tuskys forto breke *tusks*
The breste or egge or wynge or outher horn, *center, edge; other wing*
A sharre clippe hem off, right by the cheke, *shear [will] cut them off*
And with the same his wrot away be shorn, *snout*
And set it al in ordir as beforn. *it (i.e., the line)*
And if a place feynte, anoon a yawe *place [in the lines] weakens; saw (see note)*
Of myghti men aforn it is to drawe. *before*

Tribunys, erlis, or their lieutenauntys, *Tribunes, earls*
Of these, myghtiest to renne and ride *run*
Wer mad the capitayns and governauntys, *Were made*

[24] *Between every two on horseback, set one of them on foot*

And werriours hem named the subside. *warriors named them the reserve*
For thei releved th'oost on every side,
So that no man remeved from his place,
For so to doon, myght al an oost difface. *disorder*

Eek out herof thei make a boorys hed *Also out of these; boar's head*
And *cuneus* thei name it, or a wege. *wedge*
As thondirynge with leyting flammys red *flame-red lightning*
It russheth on our adversayrys egge *foe's line*
And shaketh off, ye mony a myghti segge, *mighty man says*
And if it falle on either of the hornys, *wings*
It cracketh hem, as fier to-cracketh thornys. *fire*

[Positioning Commanders (Veg. 3.18)]

fol. 36v This stood behinde al other ordynaunce;
Now is to se the place of uche estate. *each leader*
On the right honde, withoute variaunce,
The principal captayn or potestate, *commander*
That al the governaunce is taken ate, *the command of everything*
There as the foot men and the hors dyvide, *horsemen*
He hath his place, al to governe and gide. *guide*

Footmen and hors to rewle heer stondeth he, *rule here*
The potestate and al this oost to gide,
By premynence of his auctorité, *preeminence*
To chere theim that myghtily shal ride,
And theim o foote, as myghtily t'abide.
A wynge is him to bringe aboute the horn *around (i.e., flanking) the wing*
Him counteringe and on comynge beforn,

That is the lift horn of our adversary, *left wing*
Aboute a wynge, and on the backe hem clappe, *rear strike them*
And thei of their comyng the tyme wary.
And if (as God defende!) amys it happe, *a problem occurs*
Anoon the subside is to stoppe a gappe: *reserve; plug a gap*
For soveraynly on him that is t'attende, *that (the reserve) is to attend*
And, as the cas requyreth, come on ende.

The duke secounde, and next in governaunce, *second-in-command*
Amydde the frounte or forfrount is to stonde
And sustene it t'abide in ordinaunce. *encourage it to stay in formation*
The boorys hed his part is to withstonde, *boar's head*
A sharre out of the subside is at honde. *share; reserve; hand*

Clappe it theron, and if ther nede a yawe, — *there [is] needed a saw*
Out of the same anoon it is to drawe.

fol. 37r
The thridde duke, right wys and vigorous, — *third-in-command*
His part it is to stonde on the lift horn — *left wing*
And myghti men with hym, for dangerous
Is that to kepe, as writon is beforn. — *protect; written*
His wynge he muste extende, and hadde thei sworn
It, let hem not her wynge aboute hym clappe, — *(i.e., flank him)*
Subside at him be sone, if ought myshappe. — *Reserves; [sent] soon*

A clamour, clept an harrow or a shout, — *war-cry, called*
Until the fight begynne, noon is to rere. — *no one is to raise*
No werreour that wise is, out of doubt, — *veteran; without doubt*
Wil shoute afer, therwith his foo to fere. — *from afar; foe to frighten*
But when the shoute and shaftys fille his ere, — *shafts (i.e., sounds of spears and arrows); ears*
Their voyce yfere is so fel and horribil, — *voices together; fierce*
That for to fere, it is not incredibil. — *to frighten*

Be redy first, and first to sette uppon, — *ready [for battle]*
And first to shote and shoute and make affray, — *shoot; attack*
With myghti countynaunce, that is the mon, — *moon*
That mornynge is to have a ful fayr day. — *(see note)*
This promptitude and wit and stronge aray — *array*
Thi foo seynge, is trembeling to fle, — *foe seeing [it]*
The palme of victory goynge with thee. — *palm [branch]*

[Counter-actions (Veg. 3.19)]

And ay bewar, lest his right wynge clappe — *always beware; assail*
Aboute thi lift horn. This is remedie: — *left wing*
To rech it out; and if that wil not happe, — *extend*
The wynge aboute thyn horn bacward replie — *Fold back your wing in response*
And fende hem off. Now fight for the maistrye, — *fend them off*
And if a bosh come on on eny side, — *ambush come about on any*
A better bosh on hem from our subside. — *reserve*

[A Digression]

fol. 37v
Here angelike valiaunce, here is puissaunce
Archangelik in ooste and legioun,
And it governeth dukys principaunce — *governs it*
With myght, power, and dominatioun.

Omnipotens, this is His champioun!	*Omnipotence (i.e., God)*
God loveth this, His throne and sapience	*wisdom*
Is sette heron, justice to dispence.	*hereupon*
What is this oost adverse? Rebelliouns	*opposing army*
Presumptuous, perjurious, mischevous,	
Heresious with circumcelliouns!	*Heresies; vagrants*
A legioun attaynte, untaken thevous,	*attainted, uncaptured thieves*
That, as thei ar myscheved, wold myscheve us.	*led astray*
Her lord is Lucifer, the kyng of pride,	*Their*
In every feeld with him doun goth his side.	*down goes*
Thei ha no breste, here hornys and her wyngis	*center, their*
Ful febil are and out of ordynaunce.	*formation*
Subside is goon, noo socour in their kynge is,	*Reserves are gone*
And moost amonge hemself is variaunce.	*what they have most among them*
They wil away, now fle they to myschaunce.	
Goon is their herte, and if the body dwelle,	
Their hope is aftir deth and aftir helle.	*hereafter; after that*
Here is o breste, here hornys are and wyngys	*one*
And myghtieste in raunge and ordynaunce.	*[they are] the mightiest*
Subside is here, and socour in our kynge is,	
Amonge us is ther noo contrariaunce.	
We wil abide undir our governaunce,	
Here is noo drede of deth or peyne of helle.	
Here or with angelys is us to dwelle!	
fol. 38r Therfore our eye is to the kyngis signe.	*king's banner*
We here his voys, as trumpe and clarioun.	*hear*
His eyes are obeyed; we enclyne	*bow*
Attonys unto hym. His legioun	*At once*
We are, and aftir God, his regioun.	*nation*
His capitayn and his vicapitaynys	*vice-captains*
T'obey everych of us right glad and fayn is.	*To obey we are all*
This champioun, this ooste and Goddis knyght,	
With fele and also fewe may prevaile.	*many*
Miraclis here and there God sheweth myght.	
But first (as seide is erste) is hem t'assaile.	*first (as is said before); to attack them*
The gretter ooste is this. Now moste availe	
Is ordinat bataile, as is beforn	*ordinary battle*
Seide, and with wyngys clappe in eyther horn.	

[SET ACTIONS IN PITCHED BATTLES (VEG. 3.20)]

With wyngis wight hem umbego, ley on — *strong wings surround them*
Behinde and holde hem streyt on every side, — *hold them fast*
And cleche hem up. Whi wolde they be foon? — *close them; foes*
Tech hem obeyssaunce. Sey: "Fy! o pride!
Com on your way, we wil our self you gide."
This way is good, so that this bestes ride — *these rude beasts*
Be not a gret horribil multitude.

With multitude we myght been umbegoon. — *be surrounded*
War that perile; holde off on other side — *Beware; off on either*
With wyngis wight, and strengthe hem faste anoon. — *strong wings [of men]; reinforce*
With myghtiest elect of the subside — *strongest select [men]; reserves*
Prevaile on hem. Yet more is to provide,
That if the boorys hed com in, a sharre — *boar's head; blade*
Be made for him, his tuskys forto marre. — *to destroy*

fol. 38v But wurthi men are in this ooste a fewe:
Sette hem in wise and myghti governaunce!
For heer the Lord wil His myracle shewe,
Their multitude or myght be noo turbaunce. — *no trouble*
Truste in thi Lord and mak good ordynaunce.
Ordeyned wel, in fewe is to prevaile,[25]
So that theryn no poynt or poyntis faile.

Do thus when th'egys are at the congresse: — *the lines; encounter*
Thi lift hond, hold it from thin adversary, — *Your left hand*
That of his shot it have noo distresse
And thi ryght wynge uppon hem wightly cary. — *manfully*
Theer to begynne it is most necessary;
Sette on in circuyte, and bringe abowte,
And to prevaile it nedeth nat to doubte.

But do this with thin horsmen myghtyeste — *your mightiest horsemen*
And footmen of the beste, and ha noo drede, — *the best; have no fear*
Thi foomen undir foote to be keste. — *foes; cast*
And if thi foo to thee the same bede, — *to you; attempt*
A myghtiest subside uppon hym lede — *reserve; lead*
Of horsmen and footmen, and thus delude
Hir arte with arte, and theruppon conclude. — *Their artifice with artifice*

[25] *Ordered well, [with] fewer numbers [one] can prevail*

Or otherwise, if men be myghtieste
On the lift hond, the right is to retrete — *left side; retreat*
And fal on her right horn with wightieste
Footmen and hors. And til thei yelde hem, bete — *yield themselves, beat*
Hem on the bak and breeste, and overgete — *front; overcome*
Hem myghtily. But the right honde elonge, — *side stretch out*
That of thi foo noo forfeture it fonge. — *no loss is suffered*

fol. 39r War heer the boorys hed and everywhere. — *Beware here the boar's head*
Or otherwise al putte in ordynaunce:
CCCC or D pace yfere — *400 or 500 paces together*
Aforn the counteringe it is t'avaunce — *Away [from] the enemy; to advance*
Our wyngis wight uppon their ignoraunce. — *(i.e., in surprise)*
Prudence it is on hem to make affray, — *[an] attack*
Whil thei beth out of reule and of aray. — *line; array*

If hors be myghtiest, this wey is best
And doon anoon, and ellis is grete drede: — *If done at once; otherwise*
A remedy therfore is to be keste, — *attempted*
That al the light armure wightly procede,
And archerye, as sparkil out of glede. — *sparks; embers*
And embataile anoon the frounte aforn,
The breste to defende, and either horn:

If this be doon, the frounte alonge is sure.
Unlabored with fight, or otherwise,
Like as beforn is seyde, it is to cure — *remedy [the situation]*
That thi right wynge uppon his lift horn rise.
But myghtiest and wittiest dyvise
Unto that feat, and archers with hem fonge — *strike*
Of wighte men, o foote that be stronge. — *of footmen*

And this doyng, retrete thi lifte horn — *your left wing*
Fer, al abak, and raunge it like a spere, — *Far; spear (i.e., in a straight line)*
Dyvers heryn unto the way beforn, — *Different*
So that the foo noo strook theron bewere. — *foe no strike can manage*
This wil devicte anoon withoute fere. — *achieve victory*
In this manere a smal and myghti ooste
Shal overthrowe a multitude of booste. — *speedily*

fol. 39v Or finally, this ooste is but of fewe
And not so myghti men as hath the foo:
Heer hath the werreour his craft to shewe,
And embataile hym nygh a flood that goo — *near a river*
On outher half; a cragge is good also, — *On one side*

Lake or marice or castel or citee, — *marsh*
A side to defende is good to se. — *flank*

There embataile and putte ereither wynge — *both wings*
On oon side, and herwith pul off his horn, — *pull off the enemy's wing*
But fro behinde aboute is beste it brynge, — *behind around (see note)*
And with the boorys hede route in beforn. — *boar's head; ahead*
The myghtiest to this be not forborn, — *held back*
Ner they, theryn that have had exercise,
Thus hath be seyde of werryourys wise.

[ENCOURAGING THE ENEMY TO FLEE (VEG. 3.21)]

The foo peraventure is ferde and fled — *foe by chance is fearful*
Into sum holde, and ferther wolde he fle — *stronghold; farther; flee*
Fayn, wiste he how. What is the beste reed: — *Gladly, if he knew how; advice*
That he go forth, or heer beseged be? — *here be besieged*
To lete hem goon is moste utilitee
And no perile is it that foo to chace — *danger; chase*
That turneth us the bak and nat the face. — *turns his back to us*

[RETREAT AND AMBUSH (VEG. 3.22)]

Yet heer be wys and sende a fewe aforn, — *in this case be prudent; ahead*
Right aftir hem, and with a myghty honde — *stronger force*
Another way on even or amorn — *evening or morning*
Caste to come in and in their light to stonde. — *Arrange; in their path to take position*
When thei that aftir go wynne on hem londe,
Her part it is t'attempte hem esily
And so departe, aferd to bide therby.[26]

fol. 40r This seyn, thei wil suppose a wayt be goon, — *an ambush is not coming*
And disolute anoon be negligent. — *freed (i.e., from that threat) at once [will]*
Thenne is the wit, that myghti honde come on
And take hem up aslepe or vynolent. — *drunk*
Thus easily we have our owne entent, — *intention*
Therof to God the commendatioun
Be madde, and doon sacrificatioun. — *do give thanks*

[26] Lines 2144–46: *When those who went after them (i.e., the smaller force) advance on their ground (i.e., catch up to the retreating enemy), / Their job is to make a light attack / And then depart, [as if] afraid to stay there*

If part of th'ooste be fled, and part prevaile,
Heryn the prince exploye his valiaunce, — *employs*
Hem myghtily retournyng to bataile.
Forwhi? The foon be fled unto myschaunce.
Arere anoon unto your ordynaunce. — *Raise up at once your formation*
The feelde is youre, and trumpe and clarioun
And scryis make of victory resoun. — *shouts; resound*

[General Precepts of War (Veg. 3.26)]

Of knyghthode and bataile in special
Thus seide th'electioun and ordynaunce,
Here is to sette up rewlys general, — *general rules [of conduct]*
As this: The gracious good governaunce — *Such as*
Observeth everywhere. Al suffisaunce — *All needs*
Hath he that is content. Al may be born — *He has [fulfilled] who; carried*
Save wele. And scorned is that useth scorn. — *wealth*

Thi disavaile availe is to thi foo, — *Your harms help your foe*
His hurt availeth thee. Voide his advice, — *helps you*
Do thin availe. Do not as he hath do. — *has done*
In thin electioun se thou be wys.
War negligence. Do every man justice. — *Beware*
Be vigilaunt. Attende thin honour.
Thi providence be to thin oost socour.

fol. 40v Ha not to fight a knyght unexercised. — *Have; untrained*
Ha confidence in preved thing. Secré — *In secret*
Thi counsel have. Lerne of thi self disgised. — *(see note)*
The fugitif herd and untrested be. — *should be heard but not trusted*
Be gided wel by folk of that contré
That thou wilt over ride. Have in writynge
Every passage, and eke in purtreyinge. — *illustration*

Better is brede in oost to fight then lengthe. — *breadth in an army; than*
Good is in stoor to have a grete subside. — *reserve; great reserve*
With sapience socoure a feebil strength, — *wisdom support*
Fende off thi foo. Let not thin oost divide. — *your army*
Whette up thin ege. Bidde horsmen wightly ride. — *Whet (i.e., sharpen) your line*
Fight in a raunge aforn with multitude
Ayenst a fewe, and hem anoon detrude. — *subdue*

A fewer oost falle on with the right horn, — *falls upon [you]; wing*
And crokyng of the lift horn is t'elonge, — *left; to lengthen*

So that the myghtiest be sette beforn.
And if the lift horn be both wyce and stronge, *wise and strong*
Sette it beforn, and bak the right be wronge. *fight the right wing the atypical way*
Or on thin unadvised foo with wight *unaware foe*
And myghti wyngis go beforn and fight.

The light armure and every ferentary *light-armored [man]*
Aforn the frount in nede anoon procede
With subside on the wyngys for to tary. *reserves; wait*
And he that hath a litil ooste, hath nede
Of mych wit, and myghti men in dede,
And on his honde a flood or place of strengthe, *flank a river*
And either wynge on his oon horn t'enlengthe. *to lengthen*

fol. 41r Ye truste in hors: the playn is beste. Ye truste *[If] you; horsemen*
Uppon footmen: the cleef is good. Espie *hilly ground; A spy*
Amongis us to be ther is distruste:
That every man go hoom, anoon do crye,
And which is he, for with me shal espie.[27]
But sodenly this most be doon be day,
The yatis shitte, lest he go stele away. *gates shut; escape*

What is to doon, with mony take advice.
What shalbe doon, tak fewe or be alone.
Tak his advice that is secret and wyce, *wise*
Be juste, indifferent to everychone. *impartial*
For idelnesse have ay sumwhat to doone.
To straunge not, not to familier, *Not too distant, nor too*
Make of a lord. Chere a good chivaleer. *Treat; Cheer*

And here an ende I thus the thridde part *third*
In this tretice of knyghthode and bataile.
What ha we next? Forsothe, a subtil art *have; Truly*
To bilde a stronge citee, and for t'assaile
It and defense. And aftir, fight navayle,
That is bataile in ship, I here entende
For chivalers to write, and make an ende.

27 Lines 2206–07: *Order every man to go to his tent at once, / And the men will see the one who is the spy*

Part IV: Urban and Naval Warfare

fol. 41v *Ultima pars urbes parat, obsidet atque tuetur,*
Bello navali finit et ornat opus.[1]

[Proem]

This III^{de} part, as long as othre tweyne, — *third; other (first) two parts*
Halt providence of myghtiest bataile, — *Concerns the guidance*
The morthereer to bringe undir the cheyne. — *murderer; chain*
There al his olde craft shal nought availe, — *not help*
But hate of ire and angush of travaile — *hateful rage; laborious anguish*
To fynde. And aftir al that to descende
To th'everlasting deth, if he n'amende. — *does not amend [his misdeeds]*

In Brutis Albion is not to spende — *(i.e., England)*
This myghti knyghthode and bataile alone.
To Normandie and Fraunce it is t'assende, — *to rise*
Til Cristis and the kyngis foos uchone — *Christ's; king's foes all*
Be dryven out or chastised, and noone
Alyve ylefte that wil not wel beleve — *Left alive who*
And uttirly the myscreaunt myscheve. — *punish the misdoers*

Here ende I that, and to my werk releve — *return*
The laste part, anoon to bringe an ende,
And aftir in correctioun it preve. — *put it to the test of correction*
Criste truste I, that the kyng it wil attende
And werreours to knowe it condescende.
That leve I there, and write as is th'availe — *believe*
To bilde and sette assege, and see bataile. — *build; lay siege*

[1] Before 2224: *The final part prepares, besieges, and even defends cities, / Finishes with naval warfare, and adorns this work*

[Need for City Fortifications (Veg. 4.1)]

fol. 42r Nature or art assureth a citee, *artifice secures*
A dongeoun, a castel, or a tour. *tower*
In lake or in mareys or in the see *marsh*
Sette it; that element is thi socour.
And if the lond shal be propugnatour, *defender*
A mountayne or a clyef, a cragge, a rok *cliff*
Sette it uppon, and saf it is fro strok. *attack*

And in foreste, in feelde, or in champayne, *open country*
With craft or art it is to make a strengthe,
And if nature assiste, it is t'attayne
Effect anoon, as when the brede or lengthe *at once*
A rok, ryver, mareys, or see wil strengthe;
But art alone if noon herof availe, *artifice; none of these aids [you]*
Shal make it stronge with wisdam and travaile.

[Layout of Walls (Veg. 4.2)]

Mak bosumy and angulous the wal, *curvy and angled*
And so sette out therof the fundament *foundation*
With touris and turrettis overal, *towers and turrets*
That scale, engyne, or rammer therto sent *ladder, [siege] engine, or ram*
Be over sette, and faile of his entent, *overtaken*
When he is unbegon and al to-donge *surrounded; destroyed*
With al that may be kest fro wallis stronge. *cast from strong walls*

[Constructing Terraplein Walls (Veg. 4.3)]

In this manere a wal it is to make, *to be made*
To stonde an infallibil thing forever: *impregnable*
An intervalle of XX^{ti} feet be take, *20 feet is measured*
A wal on either side herof dissevre, *build separately*
Caste in the moolde, sadde it with mal and lever, *Throw; earth, compress; mallet*
Out of the dich caste it bitwix the wallys,
And ramme it doun with punchonys and mallis. *piles and mallets*

fol. 42v Mak the inner wal wel lower then withoute, *inner [part of the] wall; than the outer*
That esily, as by the clif, ascende
Me may unto the loupis al aboute, *Men; embrasures*
Or by an esi grice hem to defende. *easy flight of steps*
Thus mad a wal, the ram may nat offende.

For thaugh he fronte awey this uttir cruste,
The grounde is stronge ynough with him to juste.[2]

[Protecting Gates from Fire (Veg. 4.4)]

For firing of the yatis make obstacle, *Because burning; gates*
Cover hem with hidys and with iron plate, *hides*
And make aforn a myghti propugnacle, *build in front; defensework*
A portcolys to plumpe adoun therate *portcullis to drop down there*
Aftir thi foon. Atwixte it and the yate *Between; gate*
Thei checked ar. The machcoling may thenne *machicolations*
Chastise hem that thei shal nat sle ner brenne. *slay nor burn*

[Constructing Ditches (Veg.4.5)]

The dichis ar to make brode at al *ditches; broad at all [points]*
And deep at al, so that me may not fille *men*
Hem in no wise, and renne uppon the wal. *run [directly] against*
The myner is his labour heer to spille. *miner's work; waste*
And rathest if the watir hem fulfille: *most ideal; fills them (i.e., the ditches)*
For now hath he twey grete impedymentys; *two great impediments*
Depnesse is oon, another th'element is.

[Protecting Men on the Walls (Veg. 4.6)]

The multitude of shot is to repelle
With sheeld, pavice an here and duble say.[3]
Shot perceth not ther thorgh. Eek wittis felle *Also many wisemen*
Han cratys fild with stoon at every bay, *Have crates; wall opening*
And if th'assault come up, adoun go they
Out of the crate. At every loup is oon *embrasure*
Of these. It quelleth ordynaunce and mon. *stops formations and men*

[Insuring Food Supplies (Veg. 4.7)]

fol. 43r In mony wise assault is and defense, *many ways is; defense [undertaken]*
And on manere is by enfameyinge. *causing famine*

---

[2] Lines 2278–79: *For even if he [tears] away this outer crust of the front, / The earth [packed behind it] is strong enough to joust (i.e., fight) with him*

[3] *With shield, pavise (a kind of large shield) and hairmats and double-thickness cloth*

Hoolde foode away, and watir, kepe it thens, *keep it safe*
And hem to honde anoon shal honger bringe. *hunger*
But if we wite a seege on us comynge, *know a siege*
Anoon gete al the foode within our wonys *places [i.e., walls]*
And faste have in the multitude of stonys

Corn everydel, larder, fisch, foul, forage. *All the grain, salted meats*
And that may not be brought in, is to brenne. *And what; is to be burnt*
Wyn, aysel, herbe, and fruyt and cariage, *Wine, vinegar; baggage*
Logyng, let brenne it up, or cary it thenne. *Lodgings*
So bare it for our foon that whenne thei renne, *carry it from; foes; run (i.e., forage)*
Thei fynde nought. And use we vitaile *victuals*
With such attemperaunce, that it ne faile.

[Insuring Supplies for the Wall (Veg. 4.8)]

Glew, tar, and picch and oyle incendiary, *oil*
And sulphour herwithal to brenne engyne. *burn [siege] engines*
Charcole and cole, and al that necessary *coal*
Is forto make armure and arowys fyne
And shelde and spere, hundirdys VIII or IX, *800 or 900*
And coggys, cogulys, and pibblis rounde, *stones, cobbles; pebbles*
Fil up the wal with hem by roof and grounde. *from top to bottom*

Stoon of the flood is saddest and so best, *river; most dense*
For fourneysinge a wal and every loupe, *furnishing; loophole*
And outher with engynys to be kest *others; cast*
On hegh, adoun to falle on hed or croupe, *head or hindquarters*
Or fro the scalyng forto make hem stoupe *ladders; fall*
And have of grene tymbour grete rollys *green timber large cylinders*
And loggys leyd to route uppon her pollys. *logs; hurl upon their heads*

fol. 43v And beemys is to have of every sise *size*
And boord of every soort, and also nayl. *board; nail*
Ayenst engyne, engyne is to devise, *Against [an] engine*
And that the stuf be prest, is thin availe. *material be ready, is [to] your help*
High if it be, pulle over their topsail,
And if thei come in touris ambulary, *mobile siege towers*
Hem myghtily to mete is necessary.

[Insuring Supplies of Sinews (Veg. 4.9)]

Nerf is to have or senewis aboundaunce,	*Tendon; sinews*
The crosbowyng to stringe and bowe of brake.	*crossbow*
Hors her of mane and tail, if suffisaunce	*Horsehair*
Therof ther is, therto good is to take.	
Of wymmen here tho stryngis eke thei make:	*women's hair those strings also*
With stryngys of their her Romaynys wyvis	*Roman wives*
Saved her owne and her husbondis lyvis.	*husbands' lives*

[Insuring Water Supplies (Veg. 4.10)]

Raw hidis ar to kepe, and every horn	*hides*
The portcolis to covere, eek sheeld and targe	*portcullis; also*
And mony a thing, it may not be forborn.	*thrown out*
And if so be your watir be not large,	
To synke a welle anoon it is to charge.	*to be ordered*
For lak therof, theym that the water brynge,	
With shot defende outward and hoom comynge.	*defend [them]; coming back*
And if the welle is out of our shotinge,	*range*
Make up a tour and putte archerys there,	*tower*
For to defende tho that watir brynge.	*bring [from the distant watersource]*
Cisternys who can make, it is t'enquere.	
Make up of theym in placis everywhere.	
Rayn watir kepe in hem. When wellys faile,	*wells*
Rayn watir in cisternys may availe.	

[Insuring Salt Supplies (Veg. 4.11)]

fol. 44r A see citee this is, and salt is geson:	*[If] in a coastal city; mineral salt; scarce*
Kest watre salt in vesselling that sprede;	*Cast saltwater; spreads [it out]*
Salt wil the sonne it make in litil season.	*sun*
But thus we dar not fette it in for drede,	*But [if]; fetch*
The see gravel, gete it up in this nede,	*sea sand, gather*
Fresh watir it, and let it drie in sonne,	
And salt withoute doubte herof is wonne.	

[Defending a Wall Assault (Veg. 4.12)]

They that the wal assaulteth, bith terribil	*are a terrible*
A multitude, and trumpis proudly rynge.	

The citee nys but simpil and paisibil, *city[-folk]; peaceful*
  And ferde thei are at this first counteringe, *fearful*
And in goth they. But if the spritis springe *go they (the enemy); spirited [men]*
  And putte hem off, in comth an hardinesse, *them (the enemy)*
  And egal is fro now forth the congresse. *equal; conflict*

[NAMES OF SIEGE WEAPONS (VEG. 4.13)]

The tortoys or the snayl, the rammys grete, *tortoise; great rams*
  The sekel or the sithe and vyneyerd, *sickle; scythe; vineyard*
The cagys pluteal it is to gete *mobile screens*
  And tourys ambulary nere aferd, *mobile siege towers nearby feared*
The musculys eke with the pety berde — *mantelets also; little beard*
  Lo alle these wil this citee assaile
  With crafte, and yet with craft shal it prevaile.

[CONSTRUCTING RAMMING WEAPONS (VEG. 4.14)]

Of tymbir and of boord it is to make
  A tortoys or a shelled snail, and so
They name it. Whi? For when hem liste awake *they desire [to] awaken*
  It, out therof the hed and hornys go *[ramming] head and horns*
And in and out ayein. Oon horn or too, *One; two*
  Croked or streght, hath it, right as a snaile,
  Right as it semeth hem their moost availe. *help*

fol. 44v The bak of this tortoys, snail, or testude, *testudo*
  Wherof it hath figure and also name, *gets its shape*
With felt and heere and hidis rawe or crude, *hair-mats; raw hides*
  Lest theron fier doun cast, brenne up the frame, *burn*
Wel covered is; the sidis beth the same. *are*
  Pendaunt theryn ther goth a beem alonge,
  Therof the hed is iron-steeled stronge.[4]

Tweyne hornys if it have, it is a snaile.
  Streght may thei stonde, or the lifte horn may croke *may be crooked*
Outher the right, as may be moost availe, *Or; most helpful*
  The wal to breke and stonys out to rooke. *to tear out stones*
And if it have but oon horn, and it hooke
  A croche, it is a sikel or a sithe, *[Like] a crook; scythe*
  It breketh and out bringeth stonys swithe. *quickly*

[4] Lines 2390–91: *Hanging inside a beam runs its length, / By which the ramming head of it is strong as iron*

[Defending Against Ramming Weapons (Veg. 4.23)]

And when the frount is mad to breke and brese, *shatter*
It is a ram for that similitude,
To rush uppon the wal and al to crese *crush*
The stuf in it. Yet wil thei this delude, *they (the defenders) hinder this*
And with oon crafte thoo craftis III conclude: *those three weapons*
Of quylt and felt a trusse thei depende, *quilt; bundle*
Ther as the ram entendeth for t'offende. *Where*

Or by the hed they kecch it with a gnare *snare*
And hale it up, or by the wal endlonge, *haul*
Or turne it up-so-doun thei wil not spare.
Hem semeth it to hurte it is no wronge. *It seems to them*
And other have a wulf, this ram to fonge: *wolf; grab*
That wulf is as a payre of smythis tongys, *pair of smith's tongs*
Toothed, that in a wayt alway to honge is. *weight always to hang [down]*

fol. 45r That wulf gooth on the ram, and by the hed
Or necke anoon pulde is he up-so-doun,
Or so suspended that his myght is deed. *dead*
And other fro the wallis of the town *[the] other [defense is]*
Or out of tourys hye or of dongeoun *high towers*
Wil caste an huge ston or a pilere *pillar*
Of marbil, and so breke it al yfere. *shatter it (a ram) completely*

And if the wal be thorled therwithal, *pierced*
As happeth ofte, or doun it gooth anoon: *goes at once*
Awey with every hous, and mak a wal *make a [new] wall*
Withinne that of planke or lyme and ston; *mortar*
And if thin adversayris come uppon,
Conclude theym bitwixt the wallis tweyne, *End them between; two*
And so be quyte of this perile and peyne. *finished with; peril and danger*

[Constructing Siege Screens (Veg. 4.15)]

The vyneyerde is lighter tymburynge, *vineyard*
VIII foote brode, VI footys high, XVI *8; 6; 16*
Footys in length, and dubil covertinge *[a] double covering*
Hath it of boord and fleyk. Of twyggis grene *boards and hurdles (interwoven frame)*
The sidis are, and fier forto sustene, *to survive fire*
With felt and hidis grene it covere they, *raw hides*
So that to brenne or breke it, is no wey. *burn*

	And made ynowe of these, ar sette yfere	*[when] enough are made, they are set together*
	Unto the wal, as summe sette a vyne,	*Up to; vine*
	And tre pilers upsetting heer and there,	*setting up wooden posts*
	To make it falle, undir the wal thei myne,	*mine*
	That, puld away the stulpis VIII or IX^{ne},	*8 or 9 supports*
	Doun go the wal, this vyneyerd remeved,	*[having been] removed*
	Lest it and al ther undir be myscheved.	*all those*
fol. 45v	The cage pluteal of twiggis plat,	*mobile screens; pleated twigs*
	Of heerys hath covert and hidis grene.	*hair-mats; raw hides*
	Not over high the roof ner over flatte,	*nor too flat*
	That shot and fier suffice it to sustene.	
	On whelis III to go thei thise demene,	*3 wheels; guide*
	As goth a cart. And fele herof thei make	*many*
	With mony a wit the wallis forto awake.	*assail*

[Constructing Mantelets (Veg. 4.16)]

	The muscle shelle is but a smal engyne,	*mantelet*
	Mightily mad on whelis forto go,	
	And bere away the wallis when thei myne.	*carry; they mine [them]*
	Thei bringe stuf the dich to fille also.	
	And on the werk it may go to and fro	
	And sadde it up, that tourys ambulary	*fill it up; mobile siege towers*
	May men ynowe uppon the wallis cary.	

[Constructing Mobile Siege Towers (Veg. 4.17)]

	The muscul eke is good, the way to mende,	*mantelet also; path to construct*
	For eny thing. Of tourys ambulary,	
	To se the crafte is now to condescende,	
	Th'artificeer it nedeth not to vary.	*The constructor*
	Make hem like other housing necessary,	
	A XXX^{ti} foote or XL foote square,	*30; 40*
	And otherwhile of L^{ti} feet thei are.	*other times; 50*
	Of bemys and of boord be thei compacte,	*constructed*
	And competent, the brede hath altitude,	*in breadth as needed for height*
	With hidis grene or felt sadly coacte	*raw hides; pressed together*
	The robinge and the sidis are enclude.	*shell; covered*
	Their apparaile ashameth wallys rude,	*workings; simple walls*
	At every lyme herof ar huge whelys	*limb (i.e., axle's end); wheels*
	And brood withal the sole of every whel is.	*broad; rim*

fol. 46r Present perile is, if this tour ammoeve — *tower is moved*
Unto the wal. The place is in a doubte,
An impossibil is it off to shove, — *impossible [thing]*
Of myghtieste theryn is mony a route, — *[the] mightiest [men]; a number*
And briggis in, to renne on from withoute, — *bridges inside; run*
And scalis of al maner farsioun, — *ladders; manner of construction*
From eny part to renne on up and doun. — *any part [of the tower]*

The rammys are alongh as first engyne, — *along [the side]*
And not a fewe, a wal to overthrowe.
And undir as a vyneyerd they myne — *under [the tower]*
And briggis in the myddis are a rowe,
And fro the toppe they shote and stonys throwe.
Thus undir and above and everywhere
The wall besette. Who dar abide there? — *Who dares stay there [besieged]*

[Defending Against Siege Towers (Veg. 4.18)]

Yet here ayenst is divers medycyne. — *to defend against these*
First, if the chivalers with confidence
Go myghti out and fire this engyne, — *[set] fire [to]*
First pulde away the firys resistence. — *First [having] pulled; fire*
And if thei ha not this magnificence,
Shote at hem molliols, also fallayrys. — *malleoli; fire-spears*
But what thei ar, to knowe it necessayir is.

A malliol, a bolt of wilde fier is. — *fire-dart (malleolus); Greek fire*
A fallary, a shafte is of the same. — *fire-spear (falaricus)*
Thorgh felt and hide hem shoote: al on a fier is.
But shoote hem thorgh into the tymber frame.
With myghti alblastris go to this game, — *ballistae*
Brymston, rosyn, glewe, oyle incendiary — *rosin, glue, burning oil*
With flax doon on this shafte is necessary.

fol. 46v Or prevely with fier out of the toun — *secretly*
Over the wal, whil this tour is asclepe, — *asleep*
A feleship of fewe is let adoun, — *small company of men*
That fiere it, as noo watir may it kepe. — *set fire to*
And triced up at hoom thei skippe and lepe — *gathered up at home*
To se this ambulary touris brenne.
This hath be doon, and yet ful seelde whenne. — *has been done; yet very rarely*

[Adding Height to City Walls (Veg. 4.19)]

And otherwise is doun, the wal t'arise,
And over-go the touris altitude.[5]
Yet ther ayenst is used to devise — *against this practice*
A subtiltee, tho wallis to delude: — *to take down*
In the utter tour, an inner tour t'enclude; — *outer [siege] tower*
And when thei sette uppon this wallis blynde — *walls [with the inner siege towers] hidden*
With gabils and polifs hem overwynde. — *cables; pulleys pull them [the inner towers] up*

[Defending Against Siege Towers (Veg. 4.20)]

And beemys otherwhile, ye overlonge, — *beams; very long*
Ordeyne thei, and sette on iron hornys, — *They construct; horns (i.e., hooks)*
And as a rammys hed thei make hem honge. — *like a battering ram*
This tour with hem forbeton and throgh born is, — *batter and beaten through*
And sette ofiere, and utturly forlorn is. — *set aflame*
Yet otherwise, out of the toun a myne,
Undir the way therof, sleth this engyne. — *slays*

When this engyne on that concavitee — *excavation*
Goth with his wight uppon his myghti whelis, — *Goes; weight; wheels*
Doun goth it, into helle as it wold fle.
And this to se, the toun in joy and wele is. — *town; happiness*
But th'ooste withoute al in dolour and deel is, — *sorrow and despair*
Al desperate of help by their engyne,
And al by witty makyng of a myne. — *clever construction*

[Taking the Walls from a Tower (Veg. 4.21)]

fol. 47r But if this tour sauf sette uppon the wallis — *sits safe*
With every shot of dart, of shaft, of spere,
And dynt of axe, of swoord, billys, and mallys, — *blow; bills, and mallets*
And caste of stoon thei ley on everywhere, — *they (the besiegers) lay on*
That fro the wal awey they fle for fere. — *they (the defenders) fly away for fear*
Now to the wal, the briggis forto avale is, — *bridges are brought to help*
And mony oon goth doun anoon by scalys. — *many men go; scaling ladders*

Thei trice in other with the tollenon: — *hoist; tolleno (swing-beam)*
The tollenon a tymbir pece on ende — *[is] a wood pole on end (upright)*

[5] Lines 2504–05: *And [since] otherwise it will [fall] down, the wall [is] to be raised, / And be made higher than the tower's height*

Is sette, another twye as long theron. — *planted; twice; [balanced] on its top*
The lighter ende of it adoun thei bende.
A cageful of men therwith thei sende
Uppon the wal, when they with cordis drawe
Adoun that other ende, as is the lawe.

[New artillery (Veg.4.22)]

Sumtyme ayen this werk, the bowe of brake, — *In times before this book; crossbow*
Carribalistys, and arcubalistis, — *Carroballistae; arbalests*
Onagris and fustibulis wer take, — *Onagers (catapaults); fustibali*
And mony a dart that uncouth and unwiste is — *unfamiliar and unknown*
Amonge us heer. The taberinge of the fistis — *drumming; fists*
Uppon the bowe, and trumpyng of the gunne — *thundering of the gun*
Hath famed us as fer as shyneth sonne.[6]

Thei trumpe adoun the tourys ambulary. — *thunder down; [siege] towers*
Thei overthrowe as wel ram as tortoys. — *rams as well as tortoises*
The cage and vyneyerd therby myscary. — *screen and vineyard*
The muscul may not with his dynt and voys. — *mantelet cannot [contend]*
And countir as it goth, ther is noo choys,
But deed or quyt. For and it onys touche,
It goth for al that hangeth in the pouche.[7]

[Defending Against Attacks from Mines (Veg. 4.24–25)]

fol. 47v A conynger, that now they calle a myne, — *rabbit burrow, what*
Goth undir erth unwist. By that caverne — *Goes underground unknown*
Come in t'a toun, ye, tourmys VIII or IX^{ne}, — *into a town; 8 or 9 troops [of men]*
And prevely they rise in sum taverne
Or desolat hous, so noo wight hem werne. — *no [towns]people warn [about] them*
And sodenly by nyght uppon the yate — *gate*
They hewe, and leet their frendis in therate. — *attack; there*

And ther ayenst, if that the dwellers be — *against this; townspeople*
In touris, on the wal, or housys hye — *high houses*
Uppon the strete, is ther yit comfort? Ye! — *Yes*
So stonys out of numbir on hem flye, — *As stones beyond number; hurl*
As thaugh the buldir hailed from the skye, — *boulder*

---

6 *Has been known [to] us as far as the sun shines (i.e., everywhere)*

7 Lines 2550–52: *And contradictory as it seems, there is no choice / But death or flight. For with even one touch / It takes all that hangs in the purse (i.e., it destroys everything)*

They wil anoon retrete out at the yatis. — *at once; gates*
Now steke hem out, and stynted this debate is. — *drive; ended*

And if thei do not thus, anoon their foo — *at once their foe*
Of providence her yatis may lete stonde, — *their gates*
Until as fele as fle, wil been ago, — *many as will have fled*
And thenne in ease have hous and toun and londe.
But God defende us that we be not fonde — *found*
Aslepe so that foon lede us away
Withoute strook, or seide hem onys nay! — *said once to them no*

Lo, man, womman, and childe may keste stoon — *cast stones*
Uppon his foo from every place o lofte. — *aloft*
And therto redy sone are everychon — *ready at once*
By day and nyght; this holpen hath full ofte. — *has helped very often*
Ha stonys out of flood or feeld or crofte, — *Have [ready]; river; enclosed farmland*
Store hem on high, that in a sodeyn fere — *fright*
Fynde hem ye may, and on your foo bewere.

fol. 48r This conynger hath eek another gise:[8]
Undir the wal to crepe pryvely,
And sette up postis heer and ther by sise, — *supports here and there*
And pike away the fundament wightly, — *pick; foundation*
Ramayle it wel, the postis by and by. — *Shatter*
And when their ooste was redy, make it brenne. — *burn*
Doun goth the wall; in and uppon hem thenne!

[Defending Against Besieger Stratagems (Veg. 4.26)]

Peraventure ther is a countir-myne, — *If by chance; countermine*
So that thei faile, and feyneth a dispayre, — *feign despair*
And hem remeveth mylys VIII or IX^{ne}, — *retreat 8 or 9 miles*
Now best bewar, at market or at fayre, — *you must beware*
Or day or nyght. Thei thinketh to repayre, — *They (the besiegers); return*
If there appere among hem negligence. — *them (the defenders)*
Therfore now do grettest diligence.

Now se the wacch abide uppon the wall, — *see [that] the watch stays*
And houndis wise and grete is good to kepe. — *watch-dogs*
Eek gees is good to have in special, — *geese are*
For thei wil wake folke that ar aslepe, — *people who are asleep*
The foo comynge her welth away to repe. — *sieze*

8 *This rabbit burrow (i.e., mine) also has another purpose*

The mavlard in the dich and in the wallis, *mallard; ditch*
The martilet at scaling wont to calle is. *martin (European swift); climbing [men]*

[Defending the Siege with Breastworks (Veg. 4.28)]

The toun eke on th'assege sodenly *the besieging army*
Is wont to falle, if it be negligent.
Therfore a dich thei make up myghtily,
Without shot of everych instrument, *Out of reach*
And stake it, pale it, toure it to th'entent, *post; build it up*
Ther to be sure hemself and holde hem inne.
Thus wayteth uch on other forto wynne. *each waits on the other to win*

[Naval Warfare (Veg. 4.31)]

fol. 48v The craft t'assaulte a citee and defende
By myght and wit of knyghthode and bataile,
Honour to God, therof is mad an ende.
Now go we forth unto this fight navaile, *naval warfare*
That is fight on the see, no light travaile, *sea, no simple labor*
And not o londe. As there is so grete drede,
Therfore of governaunce hath it gret nede.

[Shipbuilding (Veg. 4.34)]

To make an hous, good stuf it is to take: *materials*
Good farsioun, and good stuf is the hous. *design*
But rather he that shippis is to make,
Se that his stuf ne be nat vicious. *inferior*
A feebil hous nys not so perilous *feeble; is not*
As is a feebil ship, other a barge.
Forthy therof the more it is to charge. *Therefore; pay attention to*

Fir and cipresse and the pynappul tre *cypress; pine tree*
Therfore is good, as seyn the bookys olde,
And ook is holden good in this cuntré. *oak*
The nayles are of bras wel better holde *brass*
Then iron. Whi? For ruste thei wil and olde *they (iron nails) will rust; age*
And kanker and consume. There as bras, *corrode; Whereas*
Consumed al the ship, is as it was. *[Even if] the whole ship is destroyed*

[WHEN TIMBER SHOULD BE CUT FOR SHIPBUILDING (VEG. 4.35–36)]

Fro Juyl Kalendis unto the Kalende — *July 1*
Of Janyveer, that is by monthis sixe — *January 1*
The seson is, tymbur to falle an ende, — *on end*
Th'umour dryinge in treen. Now sad and fixe[9]
Is every pith. But fallinge is bitwixe — *felling is [best] between*
XV and XXII^{ti}, when the mone — *the 15th and 22nd; moon*
Is wanyng; dayis VII is this to done. — *7 days*

fol. 49r In other tyme or seson if me falle, — *I fell [them]*
Worm-eton wil it ben, eek it wil rote. — *Worm-eaten; be, also; rot*
The tymbourmen of craft this knoweth alle. — *woodcutters*
Of rynde or bark is rende away the cote
And dryed thorgh, er it be put to note, — *dried through, before; use*
For tymbir weet, so wroght, wil aftir shrynke — *wet*
And ryve and with right grete disconfort drynke. — *tear apart; drink [seawater]*

For if the shippe unto the maryner
Drynke of the see, sone aftir of the same
Thei drinketh al, and are of hevy cher.
Forthi, the carpenter is wurthi blame — *Therefore*
That into shippis wil weet tymbour frame, — *wet*
And wurthi thonk is he, that frameth drye,
So that in his defaulte no men deye. — *error; die*

[TYPES OF WARSHIPS (VEG. 4.33, 37)]

The namys of the shippis as for werre
Myn auctour writeth not, save a liburne — *liburna (a small galley)*
He writeth of as myghtier and herre — *higher*
Of boord, and wight of foote, and light to turne. — *sideboard; swift*
As to the wastom of this shippis sturne, — *size; ship's stern*
Thei hadde V or IIII ordris of ooris, — *orders of oars*
Or fewer, as the vessel lesse or more is.

And every grete liburne a balynger — *balinger (i.e., a small ship)*
Hath had, and that a scafe exploratory — *scouting skiff*
Was named, forto aspie fer and neer. — *scout far and near*
Of oorys hadde that not but oon story. — *one*
But wight it was to go for a victory. — *swift*

9 *The moisture [then] drying in trees. At this time firm and fixed (i.e., stronger)*

The seyl, the maste, and every marynere *mariner*
With see colour wer clad forto unnapere. *sea-blue were covered to disappear*

[Names of the Winds (Veg. 4.38)]

fol. 49v A navey and an oost that wil governe *[whoever] will lead*
Uppon the see, him nedeth forto knowe
The wyndis. And the wedir to discerne, *weather*
He moste ha wit, leste he be overthrowe. *must have knowledge*
And first the foure cardinals arowe *cardinal directions*
Be knowe, as Est and West and North and South,
How thei amonge hemself discorde, is couth.

Th'eest cardinal is called Subsolan, *East [wind] (Latin: Subsolanus)*
And on his lifte hond hath he Sir Vulturne, *left; East-southeast (Latin: Vulturnus)*
And Colchyas is on his right hond tan, *East-northeast (Latin: Caecias); located*
Septentrion, that cardinal so storne *North [wind] (Latin: Septentrio)*
Out of the North the see wil overtorne,
Thocastias his right, and his lift side *North-northwest (Latin: Thrascias)*
Halt Aquylo, what se may theim abide. *North-northeast (Latin: Aquilo)*

Auster is cardinal meridian, *South [wind] (Latin: Auster)*
Nothus fol grymly goth on his right side, *South-southeast (Latin: Albus notus)*
And Chorus on the lifte hond forth thei han, *South-southwest (Latin: Corus)*
And Zephirus that cardinal, abide *West [wind] (Latin: Zephyrus)*
Wil in the west, and when him list to ride, *when it pleases him*
Grete Affricus shal ride on his right honde, *West-southwest (Latin: Africus)*
And Duk Favonius on his lift honde. *West-northwest (Latin: Favonius)*

If III or oon or tweyne of these up blowe, *3 or one or two*
Tethis, of hir nature that is tranquylle, *Tethys (i.e., the sea)*
Thei lene uppon, oppresse, and overthrowe,
And causeth al crye out that wold be stille.
Thei rore ayeyn, of her thei have her wille. *roar*
The shippe that this conflict seeth and hereth
(Heryn beleve me!) his hert it fereth. *heart is afraid*

fol. 50r Sum varyaunce of tyme will refreyne
Her cruelous and feers rebellioun,
Anothir helpith hem to shake her cheyne *chain*
As all the firmament shuld falle adoun
And Occian lepe over Caleys toun. *Ocean leap; Calais*
And after in a while it is tranquylle
And playne and calme, as whos seith, "Husht, be stille!"

Therfore a storme is whisedom to prevyde, *foresee*
And good it is for se serenyté, *to see serenity*
And fro the storme abide or stopp atide, *stop for a time*
And with meanabil wynd sette on the see. *[more] amenable*
Ful hard it is in peril hym to se, *to see him imperilled*
That of the wyndes had inspeccioun, *Who*
Is raysonabil in direccioun. *[And] is reasonable*

[WHEN TO SET SAIL (VEG. 4.39)]

Thenne is to se the monthis and the dayes
Of navygaunce. Forwhy? Not al the yere *navigation; year*
The wyndis on the shippis make affrays,
Sum monthis ever are of mery cheer, *Some months*
And summe loure a while, and after cleer *some frown (become overcast)*
Ynough they loke, and summe ar intractabil
And ragy wood, ancour to breke and gabil. *raging mad, anchor; cable*

The VIth kalende of Juyn, when Pliades *(i.e., 27 May); Pleiades*
Appereth — what is that? the sterrys VII — *Appears; 7 stars*
The wyndes alle ar bounden to the pees, *bound to peace*
So that ther nys noo truble undir heven, *there is no trouble*
Until the berth of Arcture al is even, *rising of Arcturus (see note)*
That is of Octobir the XVIIIth kalende. *(i.e., 14 September)*
Seecraft plesaunt hath at this day an ende.

fol. 50v Tho dayis ever are of mery cheer, *Those days (i.e., 27 May to 14 September)*
And thenne unto the III^de Ide of Novembre *(i.e., 11 November)*
The dayis wil now loure and now be cleer. *be gloomy; clear*
For unto now, as bookys me remembre,
Arcture, as from the first Ide of Septembre, *(i.e., 13 September)*
His reigne he hath, and in this meanewhile
The firmament wil loure amonge and smyle. *at times*

Novembir in tempest is al to shake,
And aftir unto Marchis Idus VI, *(i.e., 10 March)*
Viage thenne on see nys noon to take, *Voyage; sea is not to be taken*
But in the woose it is t'abide fixe. *mud (i.e., water's edge); to await*
Also by londe unused is betwixe
Alhaleweday and March to goon or ride, *All Hallow's Day (i.e., 1 November)*
But if a grete necessitee betide. *Unless*

Short is the day, the nyght is overlonge,
Thicke is the myst, and thestir is the mone, *dim is the moon*

And aftir in ther comth of wynde a thronge,	*gale*
That forto stonde he hath ynough to done,	*[just] to stand*
That is o londe. A strom is aftir sone	*Who is on land; storm*
Of leyt, of wynd, of rayn, of hail, of thondir,	*lightning*
That woful is the wight that goth therundir.

And, ovir this, in Marche, Aprile, and May,	*this [being] over*
Antiquytee of navigatioun	*Ancient authorities*
Dyvers sollennytee and grete aray	*Diverse festivals*
Was used have in high devotioun,
And eke of arte exercitatioun	*practice*
To kepe in honde. And as for feat of werre,	*hand; war*
Thei bood until the sonne ascended herre.	*waited; sun; here*

[Signs of the Weather (Veg. 4.41)]

fol. 51r And tokenys of tranquille and tempeste,	*signs*
Of wynde and rayn, thei hadden in the moone.	*moon*
Of tokenys this was surest and best:
Reed is the mone, it wil be wynde right sone,	*Red; moon; soon*
To take see theryn is good to shone.	*take sail; shun*
The pale mone is lyke to have a rayn.
The pale rede is wynde and storm, thei sayn.	*pale red [moon]*

And when the mone ariseth glad and bright,	*clear*
And namely the day that is the pryme,	*first appearance of the new moon*
Withoute humour, in hornys sharpe and light,	*clouds; crescent*
To take a grete viage is right good tyme.	*long voyage*
But if the sonne telle of eny cryme,	*crime (i.e., bad weather)*
As is if he arise undir a cloude,
That day in rayn and wynd is wont to croude.	*likely to be full*

His bright arisith is like a mery day,
His rede arisith is like a breef to blowe,	*message*
And maculous, is shour or cloudis ay,	*variegated*
And pale arisith wil reyn or ellis snowe.
A tokyn eke of rayn is the raynbowe.
In wynde and ayer, in fish and foule, Virgile	*air; fowl*
The signys seyth that may no man begile.	*Records the signs; beguile*

[THE TIDES (VEG. 4.42)]

	The maryners, thei sayn, have al this art	*mariners*
	Of wydiringe. And thei be wedir wise,	*weather-reading. If; weather*
	By discipline of it ha thei no part,	*schooling*
	But of a longe usage or exercise.	
	Wel knowe thei, the reume if it arise,	*tide (Greek, rheuma)*
	An aker is it clept, I undirstonde,	*current; called*
	Whos myght ther may no ship or wynd withstonde.	
fol. 51v	This reume in th'Occian of propur kynde	*the Ocean*
	Withoute wynde hath his commotioun.	
	The maryner therof may not be blinde,	
	But whenne and where in every regioun	
	It regneth, he moste have inspectioun.	*It occurs; be mindful of it*
	For in viage it may both hast and tary	*speed and slow*
	And unadvised therof al myscary.	*[those] ignorant of it*
	The marinere, er he come at congresse	*before he comes to strike*
	Or counturinge, uppon the see bataile,	*sea battle*
	Wil his navey so for the reume adresse,	*tide prepare*
	As may been his adverser dissavaile	*enemy*
	And hindiraunce, and also his availe.	*help*
	This may be doon anoon, for a liburne	*liburna (a small galley)*
	With wynde or oorys, as me wil, may turne.	*oars*

[NAVIGATORS (VEG. 4.43)]

The maister marynere, the governour,	*master mariner*
He knoweth every cooste in his viage	
And port saluz. And forthi grete honour	*welcome ports*
He hath, as worthi is, and therto wage.	*wages accordingly*
The depper see, the gladder he. For rage	*deeper [the] sea*
Of wynde or of bataile if ther abounde,	
The surer he, the ferre he be fro grounde.	*further he is from land*
He knoweth every rok and every race,	*rock; current*
The swolewys and the starrys, sonde and sholde,	*whirlpools; depths and shallows*
And where is deep ynough his foo to chace.	
And chese a feeld he can, bataile to holde,	*choose a field*
And myghtily sette on liburnys bolde,	*liburnae*
First with the frounte al undir see to route,	*forward underwater ram; destroy*
And, as a thought, anoon be brought aboute.	

fol. 52r The maister of the shippe, he muste be wyis.
The mariners most be ful diligent,
And myghti rowing up at point device *rowing in perfect order*
Is to been had at his commaundement,
That storne and ooris go by oon assent *stern and oars; one*
Forth right to sette uppon, and light to turne,
Ful gret avauntage haldeth this liburne.

[Tactics in Naval Warfare (Veg. 4.45)]

And as o londe an oost may be prevent *on land; fleet*
And leyde awayt uppon, right so by see *laid wait upon (i.e., ambushed); sea*
At ilis or in streytys pertynent *isles; suitable straits*
A bushement to falle uppon may be. *An ambush*
Rathest out of aray is good to se *Most especially out of order; see*
When that thei be. The reume and strem and wynde *tide; current*
With you and countour hem is good to fynde. *against*

Or wayte on hem, for wery or aslepe, *weary*
Or when thei leest of thi comynge suppose, *least*
Or in a rode as is no wey to crepe *anchorage; creep*
Away, but that ye must been in their nose. *(i.e., in front of them)*
Al that is you to wynne, is hem to lose,
And if thei can avoyde alle your cautelis, *avoid; strategems*
Thenne uch his right, the feeld and fight to dele is.[10]

Thenne in a feelde a frounte of this liburnys *these liburnae*
It is to sette, and not as on the londe *not [in straight lines]; land*
An oost. And whi? For inward it to turne is,
The hornys as a sharp cressaunt to stonde, *sharp crescent*
A bosomynge amyddis to be founde, *curving in the middle*
That umbego ye may your adversary *surround*
And close hem environ, and with you cary. *close around them*

[Naval Battle Described]

fol. 52v But on the hornys be liburnys sturne *wings; liburnae stout*
With myghtiest and booldest men of werre,
Aboute our foon of myscreaunce to turne, *our foes*
With confidence hem for to seyn: "Ye erre! *You err!*
Com undir us, and knowe your over-herre *Be obedient [to] us; overlord*

[10] *Then the field and the fight will deal to each what is rightfully his*

Moost gracioux, knowe him your souverayne.
And wil ye not? At youre perile and peyne!" *pain*

The beemys, up thei goth out of the trumpe *planks; trumpets*
And every brayn astonyeth their reson. *mind's rational thinking is paralyzed*
The firmament, lo! clariounys crumpe *curved horns*
To crye uppon, and lo! it comth adoun
With angelis, ye, mony a legioun,
To countour perjurie and myscreaunce *perjury*
And surquydrye and disobeyssaunce. *pride*

In every man thei setteth fortitude *they (the angels) bolster strength*
And high magnificence and confidence,
Perseveraunt for trouth to conclude
With adiuvaunce of myghti patience, *assistance*
And on the part adverse, an impotence *opposing party*
With couwardise and diffident dispayre *despair lacking in confidence*
Wil ferdfully with trembelyng repayre. *fearfully*

The canonys, the bumbard, and the gunne, *cannons; bombard; gun*
Thei bloweth out the voys and stonys grete, *sound of battle and large stones*
Thorgh maste and side and other be thei runne. *mast; hull; propelled*
In goth the serpentyne aftir his mete. *serpentine; target*
The colveryne is besy forto gete *culverin*
An hole into the top. And the crappaude *top [of the ship]; crapaudeau*
Wil in. The fouler eek wil have his laude. *veuglaire also; praise*

fol. 53r The covey fleeth as foulis thorgh the sayle. *volley flies as birds through the sail*
The pavice are accombred with coventys, *pavises; encumbered; projectile swarms*
Yet on thei come, and us thei wil assaile.
The bowe unnumerabil redy bent is, *innumerable bows*
The shaft fro th'ere anende it goth. Apprentys
Th'onagir is and the carribaliste,[11]
The fundubal and the manubaliste. *slinger; manuballista*

The catafract, plumbate, and scorpioun, *cataphract (see note), lead balls*
The dart and arpagoun in dayis olde *grappling-hook*
Were had. And are amonge us leyde adoun
Crosbowys yet. And crankelons ar bolde, *crank-bows*
With wildefier to brenne al in the folde: *Greek fire to burn*
The malliol goth out with the fallary, *fire-dart (malleolus); fire-shaft*
The wildefier to bere our adversary.

---

[11] Lines 2865–66: *The shaft from the ear flies continuously. An apprentice (i.e., learning its purpose there) / Is the onager and the carroballista*

Yet on they come. Awaite uppon the toppe — *top [of the ship]*
Good archery. The storm of shot as hail
So rayketh on. Thei dar not shewe her croppe — *their heads*
Ner in the mastys topp, ner undir sail. — *mast's top, nor*
Yet haile hem in a myghti voys: "Hail, hail!
Come undir your Kyng Harry! Fy! O pride!"
Thei wil not th'rof, attonys on hem ride. — *will not do it, at once*

Bende up, breke everych oore in the mytside — *Pull; break every oar; midships*
That hath a rash. Help hem, lo! thei goth undir. — *has action*
To this mysaventure hemself thei gide. — *they lead themselves*
Lo, how thei cracke on every side asondir! — *asunder*
What tempest is on hem! What leyt and thondir! — *lightning and thunder*
On grapesinge anoon, let se their fleete, — *grappling*
What hertys are in hem with us to mete! — *hearts; to meet [in battle]*

fol. 53v Armure and axe and spere of over wight — *of exceeding strength*
Is over light. As sparkelys in rede, — *Are shining bright; sparks in red [flames]*
So sparkel they on helm and herneys bright. — *harness*
The rammys and twibil the side out shrede — *rams; two-bill axe*
Of ship and mast. Doun goth the sail in dede! — *indeed*
Up goth our hook! Now it is on their gabil. — *[grappling] hook; rigging*
Lo! ther it lyeth! This batail is notabil. — *lies*

Summe into se go, fisshes forto fede, — *[the] sea; to feed*
Summe undir hacch ar falde adoun for fere, — *under hatches are fallen down*
And summe above, her hert blood to bleede, — *[remain] above decks, their heart's*
And summe seke hemself they wote ner where.[12]
And summe crye: "Alas, that we come there!
Mischefe upon mysgovernaunce betide!
Lo! pride hath us betrapped! Fy, O pride!"

"Com on! With us ye shal go se the kyng,
The gracious. Have of anoon this gere! — *Take off at once this gear*
Ye muste have on another herneysing: — *harnessing*
A gyngeling of jessis shal ye were. — *jingling of fetters; wear*
Ye shal no lenger stondyn in this fere. — *stand in this company*
O Silver Bere, O Lilial Lioun, — *Bear; Lily-white Lion*
O Goldon Eagle! Where is your renoun?"

Thus may be doon, if that it be forseyn — *foreseen*
Of our meryte in souverayn providence. — *in the sovereign's providence*
Forthi forwith do every wight his peyne — *his effort*

[12] *And some seek [refuge] they themselves know not where (see note)*

Sleuth out to holde, and have in diligence	*Sloth; diligently*
Sette up the werk. And spare noon expense	
Of Goddis honde. Although ye have victory,	
Yet in the knotte is al th'onour and glory.	*knot [of the Trinity]*
fol. 54r Knytte up the werk, and say: "Hail, Haliday!	*Holy Day*
The werre intraneous of al this londe	*internal war*
Is at an ende!" Here nys no more affray.	*is nothing more to fear*
Justice is heer peasibilly to stonde.	*peacefully*
And al the world shal telle of Engelonde	
And of the kyngis high magnificence,	*king's*
And been adred t'attempte it with offense.	*afraid*

[Naval Weapons (Veg. 4.46)]

But forto knytte a knotte uppon this book,	*compose a conclusion*
That is to sey, therof to make an ende,	
What is the ram, this twibil, and this hook,	*two-bill axe*
That helpeth us this shippis thus to shende?	*these ships; destroy*
The ram, a beem is, by the mast suspende,	*beam*
That as a saylis yerde is smal and longe,	*like a yard-arm*
On either ende an iron hed to fonge.	*fasten*
A rammys or a snailis hed theron	*ram's or snail's head*
Ther may be sette, with streght or caumber horn.	*angled horn*
On either side it may sette on our foon,	*side [of our ship]; beset; foes*
With myghti hand adoun that thei be born.	
Ther nys nothing may stonde ther beforn.	*before [it]*
For of the shippe it breketh out the side,	
Unnethe may the mast his myght abide.	*Scarcely; strength*
The hook of iron kene is and of strengthe,	*sharp iron*
And like a sithe uppon a myghti sperre,	*scythe*
And not to gret, but of an huge lengthe,	*too large*
And polissed to bace and make it herre.	*on pulleys to lower; higher*
The gabelis that in a ship of werre	*cables; war*
Bere up the sail, herwith may be fordone,	*cut*
So may the stay and shroudis everychone.	*shrouds (rigging ropes)*
fol. 54v The twibil is an axe with double bite.	*two-bill; double edge*
And therwithal in myddis of the maste	*with that, in the middle*
What maryneris dede, is hard to wite,	*mariners did; know*
But fele it hurte. And fele it made agaste.	*many; afraid*
Now faste until an ende I wil me haste,	*quickly to the conclusion*

Yet first th'onagir and carribaliste, — *onager; carroballista*
What thing it was, it were good we wiste. — *it would be good for us to know*

[NAVAL WEAPONS (VEG. 4.22)]

Th'onagir was an huge and myghti bowe,
Strynged with nerf. Therwith the stonys grete, — *sinew; great stones*
In maner of a thonderynge were throwe. — *like a thundering*
And for defaute of nerf, hors heer was gete — *if lacking sinews, horse hair*
To strynge hem with, and rather then forlete — *abandon*
The help therof, their heer Romaynys wyvis, — *the hair of Roman wives*
Kitte of, to strynge hem with, and save her lyvys. — *Cut off; lives*

Theim lever was to have her goode husbandis — *Their preference*
With honestee, and with their hedis bare,
Then dishonest be led to straunge londys,
Dispareged, her mariage forfare. — *forfeit*
O, mony oon of yon goode wyvys are,
That charge more vertue and honestee — *care more about*
Then worldly good or bodily beautee.

In carris had for hem, carribalistis — *carts built; carroballistae*
Wer sette. Thei were as bowis are of brake: — *like crossbows*
Oon more of hem then X manubalistis,
Of nerf or heer stringis for hem wer take.[13]
Their myghti shot made herte and herneys quake. — *heart; harness*
They and th'onagre bowys myghtieste: — *[are the] mightiest bows*
Tymbir that oon, stonys that other keste. — *that other (the onager) cast*

[EPILOGUE]

fol. 55r Of th'olde world the brightest herneysinge, — *ancient; harnessing*
Best ordinaunce and myghtieste mad were.
O chivalers, to you this is to bringe, — *this (the old ways)*
The beste ye chese. And yet a point go nerre:[14]
O Lady myn, Maria, lode-sterre, — *Mary, lodestar*
Licence me toward the lond. Beholde! — *Guide*
See-seke am I. Ful fayn o lande I wolde! — *Sea-sick; Very glad on land*

---

[13] Lines 2968–69: *Each of these, more than ten hand-crossbows' [worth] / Of sinew or hair-strings requires*

[14] *The best [of which for] you [to] choose. And yet a conclusion approaches*

Hail, porte saluz! With thi pleasaunt accesse, — *happy port*
Al hail Caleis! Ther wolde I faynest londe. — *Calais; most gladly land*
That may not Joon! Whi so? For thei distresse — *John (i.e., a fellow)*
Alle, or to deye, or with her wrong to stonde. — *either to die; their wrong*
That wil I not to wynne al Engelonde! — *[even] to win all England*
What myght availe? A litil heer to dwelle, — *result; here*
And world withouten ende abide in Helle!

O litil case, O povere hous, my poort — *poor house*
Saluz thou be, until that ayer amende, — *Happy; heir*
That is to sey, until another soort
Governe there, that by the kyng be sende.
Yit let me se, what way my wit is wende. — *mind is turning*
In this tretys, first is th'electioun — *the selection*
Of werreours: as for the legioun, — *legion [they should be]*

Yonge, and statured wel, of up o londe
And laborers, be taught to pace and renne — *run*
And lepe and shote, and with a dart in honde — *leap and shoot*
Shakyng uppon the Sarrasins that grenne, — *Saracens; snarl*
To shote quyk, and to swymme over whenne
The ryver is to deep. There every gise — *means*
Of hosteyinge and fight hath exercise. — *warfare; practice*

fol. 55v The part secounde hath the divisioun
Of al an oost, wheryn is tolde of th'aide, — *the auxiliary*
That subsequent is to the legioun,
Wherin t'everych office his part is leyde. — *discussed*
Theer of a feeld al ordinannce is seyde,
With evitatioun of al perile. — *avoidance*
Who redeth it, therate among wil smyle. — *reads; smile*

The III^{de} part, provideth and vitaileth — *provides; supplies*
And paeseth th'ooste, and voydeth al myschaunce, — *pacifies*
And al that in the journeyinge availeth,
Is here to rede, and what feeld may avaunce — *to [be] read*
An ooste to fighte, and every ordinaunce
How is to sette, and in conflicte how VII — *fighting how 7*
Weyis ther ar the quyckest undir heven. — *Ways*

The firthe part in crafte and in nature — *fourth*
Strengtheth a place and techeth it t'assaile, — *teaches [how] to assail it*
Engynys eek to make and putte in ure, — *put into action*
And to resiste hemself to disavaile, — *disadavantaging himself*
And on the see to make a strange bataile, — *sea; strong*

Where every feat of werre it is to spende —
And of this werk theryn is mad an ende.

Go, litil book, and humbilly beseche — *humbly beseech*
The werrioursys, and hem that wil thee rede, — *read you*
That where a fault is or impropur speche,
Thei vouchesafe amende my mysdede.
Thi writer eek, pray him to taken hede — *writer (i.e., copyist) also; heed*
Of thi cadence and kepe ortographie, — *orthography (spelling)*
That neither he take of ner multiplye. — *take away from it nor*

Finis.

Explanatory Notes

Abbreviations: **A**: Oxford, Bodleian Library, MS Ashmole 45 (Part 2), fols. 1r–7v, 18r–23v, 41r–43v, 46r–53v; **C**: London, British Library, Cotton MS Titus A.xxiii, fols. 2r–53v; ***CT***: Chaucer, *Canterbury Tales*, ed. Benson; **D&A**: *Knyghthode and Bataile*, ed. Dyboski and Arend; ***MED***: *Middle English Dictionary*; **MS**: Cambridge, Pembroke College, MS 243, fols. 1r–55v [base manuscript]; ***OED***: *Oxford English Dictionary*; **R**: London, College of Arms, MS R.25, fols. 24r–62v; ***TC***: Chaucer, *Troilus and Criseyde*, ed. Benson; **Whiting**: Whiting, *Proverbs, Sentences, and Proverbial Phrases*.

Proem

1 *Hail.* See note to line 43, below.

1–2 *kalende / Of Marche.* In their edition, D&A (pp. xvi–xvii) go to some length to try to match the poem's dating here (March 1) with the dating of their assumed occasion for composition: the Loveday between Yorkists and Lancastrians on March 25,1458. The proem, however, should almost assuredly be dated to an entry into London on March 1, 1460. The subsequent book itself, in order to be presented on that day, must have been written earlier, quite possibly in Coventry in the last week of November or the first week of December 1459 (see the Introduction, pp. 2–3). Many factors might have influenced the delay between the composition and the presentation of the book, but it could be that March, being named for the god of war, was thought a fitting time for a book on war.

2–3 *wheryn David the Confessour / Commaunded is his Kyngis court ascende.* Wakelin ("Occasion, Author, and Readers," p. 267) is correct to observe that this is not, as D&A suggest (pp. xvii–xviii), a reference to the ascension of King David to the throne of Israel, but instead a reference to the patron saint of Wales, St. David the Confessor. However, Wakelin errs in suggesting that the poet nevertheless "seems to confuse" the two figures by referring to the *kyngis court*. St. David's March 1 feast day celebrates both the life and the death of the saint (when he was called to the court of his king, i.e., God).

9–15 *And she . . . . hidir sende.* Because the "structure of the sentence is not clear," D&A term "the whole vision . . . extremely hazy" (p. 111n13–16). This implication may be due to their reading of *anende* in line 10 as *an ende*, which does indeed strain the grammar. Repaired, however, the vision seems quite clear: the poet calls for blessings upon the Virgin Mary, who comes into the procession alongside her son, accompanied by the uncountable flights of arrayed angels that they command. Nall argues that the poet goes beyond this, attempting to identify Henry VI with Christ in this imagery (*Reading and War*, pp. 125–29).

13–15 *And ordir . . . . hidir sende.* The poet foreshadows his text's integral interest in establishing a proper order for the realm — explicit in the ordering of its military forces, implicit in the unification of Yorkists and Lancastrians under the king's rule — by presenting the arrival of the angels according to their own hierarchy and under their own royal command (see note to lines 9–15, above). The tradition of the nine orders of the angels on which he relies extends back to the popular fifth-century work *De coelesti hierarchia* [*On the Celestial Hierarchy*], by Pseudo-Dionysius the Areopagite, from whom it was adopted by St. Thomas Aquinas (*Summa Theologica* 1.108). The nine orders, which will be listed in full in lines 152–58, are, in ascending order: angels, archangels, principalities, powers, virtues, dominations, thrones, cherubim, and seraphim.

17–20 *But Kyng Henry . . . . now wel.* The poet's political aims are highlighted early, as he corresponds Henry VI's entry into London with the sacred and celestial visitation just given. Just as all are made one in Christ, so, too, will all be made one through the king. Nall goes further, suggesting that the author is portraying Henry VI as a figuration of Christ, such that disobedience to the king (i.e., through rebellion) becomes heresy (*Reading and War*, pp. 130–34). Regardless, the fact that this hope came to naught is clear not just in the historical continuation of the Wars of the Roses until at least 1487 but in the poem's manuscript record, as well: subsequent references to the Lancastrian King Henry VI in the alternative manuscripts were emended to refer to the Yorkist King Edward IV (see the Introduction, pp. 11–13).

21 *Te Deum.* One of the most common Christian hymns, *Te Deum laudamus* [*Thee, O God, we praise*] was sung to give thanks to God. As a hymn that calls for God's blessing upon an event, it would be appropriate for both the royal entry into London here described and the feast day on which the event is supposedly taking place.

22–24 *Benedicta Sancta Trinitas . . . . Unitas.* The poet here incorporates the beginning of the Introit for Trinity Sunday: *Benedicta sit Sancta Trinitas, atque indivisa Unitas* [Blessed be the Holy Trinity, and undivided Unity].

25 *Deo gratias.* The traditional phrase *Deo gratias* [Thanks be to God] completes the Latin Mass, which makes it a suitable conclusion to the poet's opening vision of celestial and sacred blessings upon Henry VI's entry.

25–28 *Therof . . . . then now.* The poet here parallels the celestial Trinity with a secular trinity of the three estates of the Middle Ages: the clergy, the military, and the commons.

41 *Te Deum laudamus.* See note to line 21, above.

43 *honor, virtus.* The poet here quotes the hymn *Tantum ergo Sacramentum* [*Therefore so great a Sacrament*], written by St. Thomas Aquinas, which has liturgical use in the benediction of the Blessed Sacrament: the hymn praises the Trinity with *salus, honor, virtus* [hail, honor, virtue]. The first of these terms does not occur here, but might be implied in the "Hail" of line 1, thus neatly framing this opening of Trinitarian praise.

45–46 *Thi bille . . . . not foryete.* The poet's petition to the king appears to be his request to dedicate *Of Knyghthode and Bataile* to him, as described in the subsequent lines.

47 *my lord Beaumont.* On Beaumont, see the Introduction, pp. 2–3.

56 *Sumtyme it was the gise.* What Beaumont begins reading is line 89, the start of the translation proper.

61–80 *I wil considir . . . . exemplifying techeth.* As D&A punctuate it, Beaumont's speech that begins at line 61 does not end until the completion of the proem. To the contrary, it is here assumed that the lord's directive that the poet present the book to the king (line 64) prompts just such a presentation (lines 65–80). One cause for D&A's reading may be the placement of a period after *Offreth for ye* (in line 77), which violates a clear mark of punctuation after *Offreth* and the capitalization and rubrication of the following *For.*

79–80 *Accepte it . . . . exemplifying techeth.* One reading of these lines, Nall observes, is that the poet suggests "that the practice of war leads to God: 'mighty war leads by examples to this supreme conqueror'. If so, this is an audacious claim: the author is effectively stating that the pursuit of war leads to God" (*Reading and War*, p. 118).

PART I: RECRUITMENT AND TRAINING

89–92 *Sumtyme . . . . kingis.* It was not uncommon for writers to assert that important men did, or should, enjoy reading and contemplating ideas discussed in their own writings. Likewise, important figures often sought to portray themselves as well read and as patrons of mankind's collected wisdom. For military men in particular, this image was meant to demonstrate military knowledge and skill (C. Taylor, "Henry V, Flower of Chivalry," pp. 244–46).

93 *Jesse flour.* In Christian tradition, Jesus, by virtue of descending from Jesse's son, King David, fulfills the prophecy of Isaias 11:1: "And there shall come forth a rod out of the root of Jesse, and a flower shall rise up out of his root." Nall argues that the poet here once more makes a direct connection between the king and Christ (*Reading and War*, pp. 127–29).

swete odour. That holy individuals exuded a pleasant smell is a medieval Christian convention that could extend, as the poet uses it here, into the idea that Christian virtues are figuratively sweet fragrances. See *MED odour* (n.), sense 3.

93–165 *O Jesse flour . . . . shal fulfille.* Having at last begun translating the text of Vegetius (lines 89–92), the poet quickly diverges from his source's account of the Roman emperors in order to insert this lengthy discussion of Christian militarism. For a detailed discussion, see Nall, *Reading and War*, pp. 119–24.

131–37 *Knyghthode an ordir is . . . . patient.* The idea of knighthood as an order, especially in relation to others such as priesthood, is frequently expressed in other texts. For example, the French knight Geoffroi de Charny (ca. 1300–56), concludes one of his three chivalric manuals by repeatedly comparing the temporal and spiritual hazards faced by knights with those faced by priests, monks, and those in other religious positions (*Book of Chivalry*, ed. and trans. Kennedy, pp. 172–77, 180–91, 194–99; more generally, see also Kaeuper, *Holy Warriors*). Knightly writers invariably claim that they, themselves, faced greater peril than others, while religious writers frequently condemn the seemingly excessive behavior of knights and other military men. On the other hand, there are also instances of knights repenting of their worldly actions and committing themselves to more religious lives, such as the English knight John Clanvowe, who in 1391 wrote the penitential treatise *The Two Ways* (ed. Scattergood). On the historical context of Clanvowe's work, see J. Scattergood, "Date."

139 *Emanuel*. The birth of Jesus, according to Matthew 1:23, fulfilled a prophecy that a child would be born who would be called in Hebrew "*Emmanuel*, which being interpreted is, *God with us*."

152–58 *Folk angelik . . . . flamme of seraphyn*. Once again (compare lines 13–15, above), the poet uses the choirs of angels as comparison to his subject: this time, it is to underscore the superiority of knights over other men.

159 *man shal ben angelike*. Speaking of believers, Mark writes: "For when they shall rise again from the dead, they shall neither marry, nor be married, but are as the angels in heaven" (12:25). Here, the poet extends this notion to its seemingly logical end: men shall replace the fallen angels to fill the choirs of heaven (lines 163–65).

166–72 *Of myghty men . . . . I wil fyne*. This stanza, translating the "Ancient Synopsis" of Vegetius, presents the entirety of the poet's scope: lines 166–67 are the subject matter of the present section, Part I; lines 168–69 are the subject matter of Part II; lines 169–70 are the subject matter of Part III; and lines 171–72 are the subject matter of Part IV.

173–79 *Th'electioun of werreours . . . . an hardinesse*. Ethnography was of great interest to classical writers, including Vegetius and his predecessors. These writings, in turn, were happily received by medieval writers, who adapted and greatly expanded them for their own ends (T. Smith, "National Identity," pp. 50–66). The most widely read of these was Bartholomaeus Anglicus's *De proprietatibus rerum*, finished in the 1240s. Its massive number of surviving copies bears testament to its wide readership: two hundred complete and another seventy-three extracted, fragmentary, or heavily modified manuscripts survive of the original Latin text alone (Meyer, *Enzyklopädie des Bartholomäus Anglicus*, pp. 41–137; see corrections in Bartholomaeus Anglicus, *De proprietatibus rerum*, ed. Van den Abeele et al., pp. 18–21). In 1397 it was translated into Middle English by John Trevisa.

177 *Reserve theim to labour and to lore*. This advice does not appear in Vegetius.

194–95 *In soden case . . . . that pleasaunce*. The difficult grammar of these lines appears to be due to the poet's misunderstanding of his source: Vegetius's advice regards the need to ready city recruits if a long campaign emerges, not, as the poet translates it, a sudden one.

204 *chivalers*. This marks the first of many instances where the poet, in the course of translating the late-Antique Vegetius, introduces additional information — in this case, a preference for the youthful training of knights — that contemporizes or updates his source.

221 *No fere is it to fight*. The very real fear experienced by men in combat is rarely discussed directly in medieval writings. The most obvious example is in the Middle French poem, *Les Voeux du héron* (ca. 1346), in which John of Beaumont (d. 1356) describes how when he and other men "are in the field": "all our limbs fail us entirely, / And our enemies are approaching us, / Then we would rather be hidden in a cellar so deep / That no one could ever find us" (ed. Grigsby and Lacy, lines 369, 372–75). On fear in battle more generally, see A. Taylor, "Chivalric Conversation," and C. Taylor, "Military Courage."

227–28 *Tideus*. The reference — which is in Vegetius — is to a Homeric description of Diomedes's father, Tydeus, who is described as a man of small stature but mighty deeds (*Iliad*, trans. Alexander, 5.801). Although Homer was not read in the original Greek in late-medieval England, there were many adaptations of his stories in Latin that provided readers with different versions of his story and others during the Trojan War. However, these were not first-hand translations, but instead variants of the story through other writers. It was not until 1362 that the Italian scholar Leontius Pilatus completed his translations of Homer's two epic poems, but these did not circulate widely.

236–42 *The gentil . . . . bees gete*. The poet here very loosely translates Vegetius's quotation of Virgil's *Georgics* (trans. Fairclough, 4.92–94).

247 *Kne deep*. It is somewhat unclear what the poet is referring to here. *MED kne-depe* is an adjectival construction that means, as it does today, being knee-deep in something. This cannot apply in the current context. The equivalent passage in Vegetius says nothing about knees, though it does refer to well-shaped calves and feet that are free of fat. Given metrical constraints, the poet may be referring to knees having a "deep" cut due to the firm muscle on either side of the joint.

252 *werrys shourys*. Proverial; see Whiting S277, which cites Chaucer's *TC*: "Men sen alday, and reden ek in stories, / That after sharpe shoures ben victories" (3.1063–64).

261 *Stoor is not soor*. Proverbial; see Whiting S796.

271–77 *If chivalers . . . . in memory*. Whereas Vegetius writes broadly of how "decent birth" often begets a "suitable soldier" (*Epitome*, trans. Milner, p. 8), the poet constrains this advice of class distinctions into military roles: good breeding is a matter of concern for knights, in particular — a distinction that moves him to expound on chivalric ideas.

275–77 *And shame . . . . in memory*. Here the poet stresses the importance for knights to be brave and seek honor so that they will be remembered well. This is frequently tied to the recurring tension between bravery and pragmatism, especially in the precepts of various chivalric orders.

The Company of the Star, founded in 1351 by Jean II, king of France, is famously described by the Liège chronicler Jean le Bel:

At least once a year the king would hold a plenary court which all the companions would attend, and where each would recount all the adventures — the shameful as well as the glorious — that had befallen him since he'd last been at the noble court; and the king would appoint two or three clerks who would listen to these adventures and record them all in a book, so that they could annually be brought before the companions to decide which had been most worthy, that the most deserving might be honoured. . . . Moreover, they had to vow never knowingly to retreat more than four *arpents* [ca. 280 meters] from a battle: they would either fight to the death or yield as prisoners. (*True Chronicles*, trans. Bryant, p. 217)

However, after a few years most of the order were killed at the battles of Mauron (1352) and Poitiers (1356), and the Company was disbanded shortly thereafter (Boulton, *Knights of the Crown*, pp. 174–88).

279 *wagys*. D&A regard the poet as here in "error" for having utilized wages where Vegetius refers to "years of service" (p. 115n279). As Nall observes, however, this is almost assuredly a case in which "the translator was updating and recontextualising *De re militari* for a fifteenth-century audience, rather than mistranslating and mindlessly adding words for the sake of rhyme" ("Perceptions of Financial Mismanagement," p. 124; *Reading and War*, pp. 53–54). The poet's concerns here thus reflect broader contemporary concerns about the roles of wages, supplies, and discipline in diagnosing English defeat in the Hundred Years War (see the Introduction).

285 *Civilians*. Here, the term means "an authority on or administrator of civil law" (*MED civilian,–ien*). It was not until the modern period that 'civilian' took on its commonly-understood meaning of somebody not involved in war.

292–93 *knyghthode . . . . bacheler elect*. Here and throughout the poet uses *bacheler* to mean 'recruit' rather than the knightly position of knight bachelor, a lower ranked knight who fought under the banner of another knight, typically a knight banneret, who had his own banner. See also note to lines 1853–54, below.

300–01 *Conflicte is not so sure in multitude / As in the myght*. The idea of numbers not mattering as much as other factors, especially bravery or divine support, is a biblical trope (see 1 Machabees 3:18–19) commonly expressed in medieval descriptions of war.

307 *How may a thing informal fourme me*. Building on the previous line — a fairly direct translation of Vegetius — the poet here reveals his erudition: in Middle English, a thing informal lacks form; that is, it fails to accord with the rules of logic, one of the three fundamental arts of the medieval education (the Trivium: grammar, logic, rhetoric). The poet thus wonders, in a rhetorical flourish (rhetoric being a second aspect of the Trivium), how something illogical could provide substance . . . a clever play, in turn, on the theological underpinnings of God providing form — both physical and logical — to Creation.

311–12 *Celsus Cornelius . . . . Vegetius.* Celsus Cornelius (25 BCE–50 CE) is famed for an encyclopedic work of which only the medicinal book survives. Cato the Elder (234–149 BCE), general and statesman, wrote one of the first histories of Rome: his now-fragmentary art of war, *De re militari,* was heavily used by Vegetius. Vegetius also listed as authorities Frontinus and Paternus (secretary to Marcus Aurelius), as well as the imperial constitutions of Augustus, Trajan, and Hadrian. That Vegetius should have his own name listed among those whose works he accorded such honor would no doubt please him.

319 *the good angel us lede.* The poet is likely referring to angelic guidance in general; the image, after all, is a popular one in Christian stories. Within the Bible itself, it is an angel who brings word to the Holy Family that they must flee to Egypt to escape Herod's massacre of the innocents (Matthew 2:13–21).

325–26 *And XX* ML *pace . . . . over blyve.* Roman practice was to divide the daylight hours of a given day into twelve equal parts, thus making summer hours longer than winter hours. In describing the Roman military pace, Vegetius specified that the hours in question were summer hours, which equate to approximately six hours by our standardized divisions of time today. Thus the Roman military pace being described was roughly 3.33 mph, quite close to the 3.4–3.5 mph pacing that is standard marching pace in most modern militaries today.

332 *to serche or sture.* Where Vegetius remarks specifically on the speed of runners in both coming and going from scouting, the poet allows serche to cover all, while adding that runners also have a use in "stirring" — likely a reference to acts of incitement.

334 *Rennynge is also right good at the chace.* While true, it was far more effective to use men on horseback to chase down the enemy as they fled the battlefield. Battles in the later Middle Ages were mostly fought on foot, although leaders sometimes kept reserves of horsemen or had horses at the ready to mount and pursue the defeated enemy.

342 *brigge.* Bridges were of great strategic importance, especially in spring when rivers swelled. Due to their equipment and supplies, armies could only cross rivers with the use of bridges or after finding an appropriate location to ford, both of which would have been well known to defending forces and thus well guarded. Among the many occasions highlighting the importance of bridges on campaign, the 1346 Crécy campaign of Edward III is one of the most famed (see Livingston, "Losses Uncountable," pp. 8–11).

346–47 *Knyght . . . . is to leve.* The importance of having supporting members of the army, such as cooks, able to keep up with the advance of the main force is best illustrated in the example of Edward III's 1327 Weardale campaign, in which a portion of the English army advanced to try to find the Scots, but left nearly all of their supplies and other supporting men behind them. Their misery is described in extensive detail by Jean le Bel, who claims to have accompanied the English army on the campaign (*True Chronicles*, trans. Bryant, pp. 38–44).

357 *as wel swordmen as knyghtys.* D&A consider swordmen to refer to Vegetius's *gladiatores,* and that the poet thereby "breaks his rule of omitting details specifically Roman" (p. 117n357),

but the swordsmen referred to are far more likely to be contemporary sword-armed footmen (a *gladius* being a sword, after all) than arena fighters.

371 *Seynt George.* During the late Middle Ages, St. George, a Roman soldier martyred for his faith, grew increasingly popular in England. He would become the patron saint of England during the English Reformation, but here, in this far earlier text, his importance as a national figure is already clear.

375 *It is the Turk.* D&A (pp. 117–18n373–375) note that it was a fairly common practice in the Middle Ages to dress up training pillars to appear as Saracens.

398 *doctour.* A *doctour,* in Middle English, most typically means an authority, particularly one in theological matters, but extending down through the arts and sciences. The poet's use of the term to apply to a drill instructor is due to Vegetius's use of the Latin term *campidoctor* (drillmaster).

399–403 *For double wage . . . . in armys fight.* D&A regard the poet's translation as a "misrepresentation of the original facts" (pp. 118–19n399), but it is hardly so mistaken as they surmise: only the specificity of inadequate trainees needing to prove themselves *tymes VIII or IX* (line 401) is lacking in Vegetius, who simply says multiple demonstrations were necessary.

401 *prince.* The poet collapses Vegetius's ranks of Roman officers into a more general authority figure.

402 *whete . . . barly.* This differentiation of grains, which comes from Vegetius, reflects Roman agricultural production and eating habits. Wheat was the preferred grain to eat, but barley was a hardier crop that could be grown across a wider range of soils and was therefore much cheaper to buy.

404 *Res publica right commendabil is.* Classical discussions of such topics as governance and war were popular in fifteenth-century England, especially thanks to a wider availability of Roman writings, both in Latin and in English and French translations. These led writers to discuss alternative ideas of government and the role of knights and other warriors in society more openly. Such ideas were particularly influential in France (Allmand, "Changing Views of the Soldier").

411 *Caton.* I.e., Cato the Elder. See note to lines 311–12, above.

420 *artilaunce. MED,* with the sole citation of this line in D&A, provides the vague definition of "?Arms, the use of arms." Given the spelling and context of the term, however, it is likely that the poet has in mind a neologism regarding the art of using the *launce,* a term that could mean equally a horseman's spear (a *lance* in Modern English) or one that is thrown (often termed a *javelin*).

431 *The werreours thus taught, shal make peax.* D&A regard this addition by the poet as a "transition from the general body of soldiers, who, having acquired these accomplishments, may 'rest in peace' (a very unmilitary expression), to the one-third or one-fourth selected for further training" (p. 119n431). This appears to be the thinking behind *MED pes* (n., sense 5e), which cites this line meaning "to conclude one's military training." However, one should not neglect the greater vision of the poet that warriors sufficiently trained — and wars thereby sufficiently won — are meant to bring about a sufficient peace. This idea that peace might be had through war was commonly expressed in other writings on war, especially English narrative sources (T. Smith, "National Identity," pp. 131–36). However, battles were rarely decisive. Wars in the later Middle Ages were fought on a large scale and were anything but brief.

440–41 *if so be thou ride, / On hors is eek the bowys bigge uphale.* English archers, though they could ride horses on campaign, dismounted to fight. Archers fighting from horse were more the purview of eastern forces, most notably those encountered by crusaders.

442 *Smyte in the . . . bak.* The poet's instruction that it is acceptable to strike your enemy in the back appears to go against ideas of what it was to be an honorable knight. However, ideas of chivalry were quite fluid and difficult to pin down in the Middle Ages. There does not seem to have been any strict rules on where you could or could not attack your enemy in battle, so long as you were not breaking some sort of pre-existing agreement or oath on what you would or would not do. Rules in jousts were more restrictive, however.

448 *Caton.* I.e., Cato the Elder. See note to lines 311–12, above.

450 *Claudius.* Appius Claudius Pulcher (d. 211 BCE), who served as Roman consul in 212 BCE, was an active, if not always successful, commander in the Second Punic War. He was one of the two commanders during the siege of Capua, at which light-infantry *velites* (dart-throwing skirmishers) were first utilized. Vegetius gives Pulcher credit for this development, though most writers give the honor to his fellow consul at the siege, Quintus Fluvius Flaccus.

451 *Affricanus Scipio.* Publius Cornelius Scipio Africanus (236–183 BCE) was a Roman general and consul (205–04 BCE), best-known for defeating Hannibal at the battle of Zama (202 BCE), for which Scipio was given the agnomen of Africanus.

457 *breketh ofte and breseth flesh and bonys. MED brisen,* sense 1a cites this line's *breseth* under the meaning "shatters," but flesh does not fracture. It should be listed as sense 5, meaning "bruises," thus sensibly giving the line as "often breaks bones and bruises flesh."

463 *caste stoon.* It is not entirely clear what sort of "casting" of stones the poet imagines here, whether through slings, by hand, or tossing down from city or castle walls.

469 *grete estatys.* The poet here truncates Vegetius's discussion of ancient Illyricum, which was said to have had two legions that used these weapons to great effect.

472–73 *Dioclisian / And his co-emperour, Maxymyan.* Gaius Aurelius Valerius Diocletianus and Marcus Aurelius Valerius Maximianus were co-emperors of the Roman Empire (r. 286–305), with the former in charge of the east and the latter the west. They are notable for persecuting Christians, although it appears that Maximian only did so at the instruction of Diocletian.

474–79 *The chivalers . . . . sword and shelde.* Running around and leaping onto a horse from the ground while in full armor is not as far-fetched as it sounds, as has been demonstrated by the rigorous practical experiments by Daniel Jaquet et al. ("Range of Motion").

488–94 *And exercise him . . . . dar wel fight.* The poet deems to skip much of the next chapter of his source (Veg. 1.20), which is a long discussion of historical arms and armaments that had fallen out of use even in Vegetius's time. In its place, the poet provides advice about keeping battle-gear in good condition before picking up Vegetius's final words about the effect of armaments on courage (lines 493–94).

488–89 *And exercise him . . . . to were.* Contrary to the popular image of knights in full plate plodding along with little mobility, a properly fitted suit of armor would allow great freedom of movement. The poet here advocates men training in their armor so that they can be better used to bearing the extra weight while on campaign and, most importantly, in combat. Other writers occasionally describe such practice, most notably the anonymous writer of the biography of Jean II Le Maingre, Boucicaut (1366–1421). See note to lines 474–79, above.

500–01 *For faute of wacch . . . . not preved.* D&A rightly note that this couplet "may allude to some skirmish in the wars of the Roses," but they err in stating that "it is impossible to identify the occasion" (p. 121n500–501). There is every reason to suspect that the event referenced is in fact the First Battle of St. Albans, fought on May 22, 1455, just a few short years prior to the poem's assumed date of composition. In the engagement, the assaulting Yorkists attacked Lancastrian positions within the town with such speed that they caught many of the defenders unprepared for battle: according to the contemporary account of the "Phillipps Relation," Sir Robert Ogle and his men had already seized the market when "the larum [alarm] belle was ronge, and every man yed to harneys, for at that tyme every man was out of ther aray, and they joynid batayle anon" (Ed. Gairdner, *Paston Letters* 3.30), and at least one report of the battle, popularized by the London chronicles, even accuses the Yorkist earl of Warwick of commencing his assault while negotiations were still in progress (Armstrong, "Politics and the Battle of St. Albans," p. 39). The Lancastrians were routed, King Henry VI was captured, and the Wars of the Roses were begun in earnest. Such a lack of preparedness could well be termed both a failure of the watch (the poet's present subject) and a reason why being consistently harnessed for battle was necessary (the poet's previous subject).

502–08 *To make . . . . therfro.* Armies in the Middle Ages often constructed temporary defensive structures while besieging an enemy fortification or encamping in hostile territory, although these were rarely to the sort of scale advocated by the poet here.

504 *watir deep myd-thigh.* This addition to Vegetius, described by D&A as "for rhyme's sake merely" (p. 121n504), is to the contrary quite pertinent to real-world logistics. A water source

must be of sufficient depth both to have enough volume for the sustenance of the army and to ensure that it is not immediately tainted by its use.

519 *estward.* Orienting the camp *estward* by default appears to be a Christian emendation to Roman practices added by Vegetius. See also note to lines 528–29, below.

523–24 *centenaryes .... dragonys.* The dragon was commonly employed by the Romans as a standard, although perhaps not to the extent that Vegetius suggests. It clearly resonated with the poet and other English readers, as Geoffrey of Monmouth, in his *De gestis Britonum* (ca. 1095–ca. 1155), states that it is the standard of King Arthur and his father, Uther Pendragon, who is himself inspired by Merlin's prophecies, in which dragons play a prominent role (ed. Reeve, §§ 111–13, 116, 132–33, 135, 147, 164, 168, 171).

525 *Gorgona the wicche.* Gorgons were mythical creatures frequently portrayed in Greek and Roman literature. They are often depicted with hair made of snakes. Those who looked upon them were turned to stone. The most famous of these was Medusa, who is killed by Perseus in various stories. Despite its popularity in stories, the gorgon was not used as a common heraldic charge in the high and late Middle Ages. The poet might be contrasting the gorgon, as a defensive symbol (they defeat those who attack them by turning them to stone), with the dragon mentioned above, an offensive symbol (they fly around, attack people, and seize treasure).

528–29 *Ther trespassers .... north, or westward.* The poet seems to have been confused by an admittedly confusing passage. Roman practice, according to Vegetius, was to arrange the camp with its main gate pointed north, or toward the enemy or the line of march. Christian influence shifted this north-south axis to an east-west axis, with the main gate now pointed east: the direction of the rising sun and, roughly speaking, Jerusalem. Vegetius is clear that any delinquent soldiers were taken out the rear gate for their punishment, which would be either the gate to the west (in a Christian-influenced practice) or south (in the traditional practice). Between all this talk of compass points, the poet ended up with the traditional rear gate on the opposite side from where it should be.

530–37 *In III maneer a strengthe is to be walled .... a dike tumultuary.* While the poet recognizes that Vegetius says there are three different ways to fortify a camp, he grows confused on how they relate. The first, according to Vegetius, is a simple 3-foot trench, with the turf piled to one side as a simple barrier (lines 532–34). The second is a more extensive 9-feet broad, 7-feet deep trench with the earth piled up accordingly; the poet mistakenly adds this to the first to create a 10-foot deep earthwork (lines 534–37). The third option is the ditch that is 12-feet broad and 9-feet deep.

547–48 *That stake of pith which every man himselve / Hath born.* Some men in armies occasionally carried stakes of wood with them for temporary defense works. The most famous example of this is the English archers at Agincourt (1415), who deployed their pointed stakes in front of them to defend against the charges of the mounted French knights.

572–78 *We werreours . . . . to prevaile.* Of note, the poet here moves into first-person as he describes the actions of the warriors in battle. As Nall points out, here and elsewhere the first-person is used to "emphasise the author's solidarity with the Lancastrian cause" (*Reading and War*, pp. 136). It does not, however, indicate that the writer was once a combatant (D&A, p. xxxii n2). See also lines 1633–41 and 2637.

578 *undir grate. MED grate* (n.3), sense1c cites this line as the sole occurrence of the meaning "?a breastplate," extending this sense (without explanation) from the word *graper*, which was the leather grip on the back of a lance that helped a knight stabilize the weapon against his side. Even if a *grate* could be a *graper* but really mean a *breastplate* — nowhere else is this so — such a usage would fail to make sense in the discussion of field formations here unfolding. To the contrary, *MED graten* (v.), sense 1c, meaning an attack, fits the context quite perfectly.

598 *Bowed, tacled, darted, jacked, saladed.* This string of participles, referring to bows, arrows, darts, padded jack, and sallets, summarizes the principle that men should be equipped with their proper equipment while marching to stay fit. It does not mean that all men, even men-at-arms, would be armed with all of this equipment.

614–20 *Th'electioun . . . . not to tary.* The poet skips over the final chapter of Vegetius (1.28), which concludes Part I with a discussion of Roman valor and a repeated call for the adequate training of forces of arms.

Part II: The Military Structure

626–27 *To rede a thinge withoute intelligence, / As seith Cato the wise, is negligence.* The Latin form of this proverb, *Legere et non intelligere neglegare est*, concludes the preface to the *Distichs of Cato* (ed. and trans. Chase, p. 12), a collection of sayings on wisdom and morality. Written in the third or fourth century by Dionysius Cato (not the more famous Cato the Elder, as often assumed), the *Distichs of Cato* was enormously popular throughout the Middle Ages and well into the eighteenth century.

633–34 *As myghti herte . . . . metris springe.* This statement on knights appreciating poetry has many echoes in courtly poetry of the twelfth and thirteenth centuries.

640 *Crede.* That is, the Apostles' Creed, a brief statement of Christian faith that, by tradition, goes back to the Twelve Apostles of Jesus.

642–43 The reference is to the first line of Vergil's *Aeneid*: *arma virumque cano* (I sing [of] arms and a man).

651–53 *In too maner . . . . up and doun.* The two kinds of vessels, according to Vegetius, are *liburnae* and *lusoriae*, but which English equivalent terms the poet thinks are so well known that they need not be mentioned is unknown to us now. D&A suggest *barge* and *balinger*, which are possible but by no means certain (p. 123n652).

660–62 *Confederat men . . . . we in Fraunce.* Both in Roman times and in the Middle Ages, it was a widespread practice to increase the size of the army by employing mercenaries, sometimes from quite foreign shores indeed. Here, the poet remarks on how this could mean Frenchmen fighting for the English and vice versa.

672 *diagalange.* This term typically means a type of medicine, made primarily from galangal and sugar, that produced warmth and was employed to treat a variety of symptoms, including general illnesses, epilepsy, dropsy, and other excesses of cold humors (Ed. Norri, *Dictionary of Medical Vocabulary,* 1:288). The term is used figuratively here to mean superfluous detail (s. v. *diagalange*; see also *MED diagalange*). Writers were familiar with such medical terms, as they were found in a number of sources, especially in herbals and collections of medical recipes, that were commonly included in unrelated manuscripts.

673 *not worth a pulled hen.* Proverbial; see Whiting H347, and compare Chaucer, whose Monk so disregarded the Benedictine Rule that "He yaf nat of that text a pulled hen" (*CT* I[A]177).

716 *chivalers . . . the flour.* It is a late-medieval commonplace to refer to the best example of something as its "flower," but referring to the finest knights as the flower of chivalry or knighthood became particularly popular during the Hundred Years War (see *MED flour* [n.1], senses 5a, 5c). See, e.g., Chaucer's Knight's Tale, in which Arcite is referred to as "of chivalrie flour" (*CT* I[A]3059).

718 *VIIIXX and VI of wight horsmen.* 'Eight twenties and six' (=166) cavalrymen differs from the expected count of 136 (the total for the entire legion being 730, as the poet confirms in line 749). As D&A rightly observe, manuscripts of Vegetius vacillate between 132 and 136, but none agree with the poet's number here (p. 124n718).

749 *VII C . . . and XXX.* The best Latin copies of Vegetius, as used by Reeve in his critical edition, record 730 here, as the poet does. Milner's translation, however, follows a different family of manuscripts that record 726.

782–88 *The clarioner . . . . it prevaile.* Vegetius names three different kinds of horns — *tuba, cornu,* and *buccina* — along with three corresponding players for them, and the poet gamely tries to translate these Latin words into English, but it seems clear that he was unsure exactly what Vegetius intended. Modern scholars are in no better position. How these instruments ought to be defined is much disputed, including whether there were even three different instruments in the first place. Our glosses try to reflect this vague differentiation. On the general concept of using musical instruments for signaling in medieval battles, see Contamine, "La musique militaire."

797–98 *and so to IIII and V, / That hadde wage, uche aftir his estate.* Vegetius lists single, double, and pay-and-a-half for various ranks of *torquati*; the poet's furthering of the principle here may have more to do with rhyme schemes than reality.

802 *the souldeours, uch othrys mate.* The poet appears to collapse the final rank of the officer class, Vegetius's *munera,* with the common soldiery.

803 *principal prince of the legioun.* Vegetius refers here to the *primus princeps,* typically meaning the most veteran of the centurions.

845 *Not LX.* The various figures given above suggest that this was supposed to be 50, not 60, and that *LX* is a scribal error followed also in C and R (leaf missing in A).

852 *consulys.* Whether for poetic reasons or out of ignorance, the poet fails to note that the legates sent by the emperor are ex-consuls, not current ones.

869 *in payne or deth t'anende.* It was becoming more common to have ordinances for English armies from the end of the fourteenth century onwards. These listed illicit behavior and the sorts of punishment that could be expected. In some cases, such as desertion, pillaging, and killing fellow soldiers, men could be executed. See Curry, "Military Ordinances."

927 *by rowe.* That is, the ensigns for each century had markings to indicate both the cohort and the placement of the men within it.

930–31 *in their salet / That had a creste of fetherys or lik flourys.* The poet is here using *sallet* to mean a general type of helmet, not the specific variety that we might use the word for, in order to fit the meter.

946 *spende his wepon wel withoute waste.* Where Vegetius observes that a good centurion is properly trained in a set of martial drills, the poet translates this theoretical art into its practical effects: a warrior relying on trained skills is more efficient in his actions.

977–78 The poet omits Vegetius's chapters 15–25, a large portion of Book 2. D&A suggest that this may be due to "an accidental gap" in the surviving manuscript record, though the initial line of Book 3 "would rather seem to point to intentional condensation" (p. 129n977). The omitted chapters provide details on the ancient legion: its battle-order (chapters 15–18) and administration (chapters 19–25). The poet may well have regarded this material, which was not in full use even in the time of Vegetius, as having particularly little contemporary value.

PART III: FIELD TACTICS

981–82 *The salvature . . . . every regioun.* D&A attempt to tie *religioun* and *regioun* to "the oaths of allegiance" that Vegetius presents as necessary to membership in the Roman armies (p. 130n981–2). More likely, especially in light of the subsequent digression on a battle from the Wars of the Roses, the poet is here speaking of contemporary applications of his translation of Vegetius.

984 *by the werre His reste to purchase.* See note to line 431, above.

985–1026 *O gracious . . . . knyghthode and bataile.* This account of the present status of the Yorkists relative to the poet's own Lancastrian cause may well indicate the date of its composition: it clearly references the aftermath of the Battle of Ludford Bridge, a Lancastrian victory over Yorkist forces that took place on October 12, 1459. The fact that the poet does not take the opportunity here to laud his king for an even greater victory against the Yorkists at Wakefield on December 30, 1460 indicates to D&A that "the date of this part of the poem appears fixed between October, 1459, and December, 1460" (p. 130n984–1026). In point of fact, the poet's apparent ignorance of the Yorkist victory in the Battle of Northampton on July 10, 1460 — in which his addressee, King Henry VI, was captured — may further narrow this likely date of composition.

986 *Summe are in Irelonde.* Richard, duke of York, after his defeat at Ludford Bridge, fled to Ireland with his second son, Edmund, earl of Rutland. York and his son had previously held the titles of Lord Lieutenant and Lord Chancellor of Ireland, respectively, and the parliament there remained loyal to their interests.

987–88 *In Walys other are, in myghti place / And other han Caleys with hem to stonde.* After the battle, Richard Neville, earl of Salisbury, and his son, the earl of Warwick, joined by York's son, Edward, earl of March, initially fled to Warwick's estates in South Wales. From there, the party journeyed south, and on November 2 arrived in Calais, which had held loyal to Warwick's position as captain of Calais under the leadership of Warwick's uncle, Lord Fauconberg. They arrived just ahead of the duke of Somerset, newly appointed by Henry VI to replace Warwick as captain, setting off a lengthy stand-off. On the question of whether the journey of Edward and the Nevilles to Calais was a storm-driven accident or a dynastic strategem — surviving chronicles differ — see Jones, "Edward IV, the Earl of Warwick," pp. 343–44.

989–91 *Thei robbeth . . . . sle withoute pitee.* Between November 1459 and January 1460 two fleets carrying men, supplies, and arms intended to support Somerset's attempt to wrest control of Calais from Warwick were captured by the Yorkists. These comments about the Yorkists raiding and killing, in a rather unethical fashion, engage with the greater debates on the role of military forces on either side in protecting, or undermining, the common wealth (see Nall, *Reading and War*, pp. 160–61).

992–1026 *The Goldon Eagle . . . . knyghthode and bataile.* The use of animals to represent different important persons, especially kings, is particularly common in medieval political prophecies. Some of the more popular examples of these are the versions of the *Prophecy of the Six Kings to Follow John*, ultimately based on the prophecies included in Geoffrey of Monmouth's *De gestis Britonum* (ca. 1095–ca. 1155), and the *Bridlington Prophecies* (1349–50). The use of animals rather than names or titles added character and imagery to what might otherwise be terse descriptions, while also giving the texts further authority by remaining a bit inscrutable and, maybe, "magical." Their peculiarly imprecise descriptions also allowed these prophecies to be adapted to suit later events and attitudes. Here, the poet doubles this prophetic practice with the heraldic practice of identifying nobility by the animals on their coats-of-arms.

992–93 *The Goldon Eagle and his Briddys III, / Her bellys ha they broke, and jessys lorne.* The Eagle, as it does in Shakespeare (*Henry VI, part III*, ed. Evans, Tobin, and Baker, 1.1.268, 2.1.91, and 5.2.12), represents Richard of York. The three Birds must then be his three eldest sons, Edward, Richard, and Clarence. The poet here conflates falconry — the three birds have the requisite bells (*bellys*) and fetters (*jessys*) — with this popular imagery.

994 *The Silver Bere.* Richard, earl of Warwick.

995 *bare is he behinde and eke beforne.* In describing the earl of Warwick as bare behind and in front, the poet makes his dislike of the man abundantly clear by conflating a number of provocative images and statements. The earl, in his view, could be construed as bare in a moral sense (for having worked against the king) and bare in a financial and political sense: parliament had responded by stripping the earl of his lands and title.

996 *The Lily-whit Lyoun.* As D&A note (p. 132n996), this iconography would fit Edward, earl of March, but he was already accorded a place in this heraldic allegory as one of the three Birds with York's "Golden Eagle" (line 992). The next likely candidate, then, would be John Mowbray, duke of Norfolk. Mowbray had been an early Yorkist but moved toward the Lancastrian position as the Nevilles gained in influence and took no part in the events at Ludford Bridge on October 12. A member of the so-called Parliament of Devils that began in Coventry on November 20 and passed bills of attainder for high treason against the Yorkist leaders, Mowbray joined the other attending lords in swearing an oath to the Lancastrian succession on December 11. If, as seems likely, Mowbray is indeed meant, then the composition of this part of the poem might be further constrained to the immediate weeks following October 12, before Mowbray was clearly sworn to the Lancastrian cause. Alternatively, it could be that the poem was composed subsequent to Mowbray's oath and marked continued suspicions regarding the duke's loyalties. If so, the suspicions were well-founded: after the Yorkist victory at Northampton on July 10, 1460 Mowbray once more turned to open support of their cause.

1000 *Antilope.* That the poet observes how the good nobles properly attend on the Antelope — that is, King Henry VI — might also point to composition related to the Coventry Parliament of 1459. See note to lines 1014–19, below.

1014–19 *Goode Antilop . . . . fild with galle.* Within the allegory here presented, Henry VI is presented, through one of the king's heraldic badges, first as the Antelope, who desires only peace. The poet next allies the king to Christ via the Panther, another of the king's heraldic badges: medieval bestiaries often record that the panther's sweet breath entrances all animals but the dragon (Ed. Morris, *A Bestiary*, pp. 24–25); due to the latter beast's obvious association with Satan, the panther became a symbol for Christ in the Middle Ages. Those who oppose the Antelope and Panther are thereby likened to the Dragon. Beyond the dragon's likeness to Satan — and imagery of its defeat at the hands of Saint George, the patron saint of England — it could less obviously point to the Welsh who supported the Yorkist earl of Warwick and had harbored him for a time after his flight from Ludford Bridge (see note to lines 987–88, above). Similarly, the dragon being *filled with galle* points to its bitterness, while also potentially incorporating wordplay with Gaul (i.e., France), where the Yorkist stronghold of Calais was found. Here, as elsewhere (see note to lines 17–20, above), Nall goes further in more fully conflating Henry VI with Christ (*Reading and War*, pp. 128–30). Regardless of that possible association, it should be noted that, since the poet has consistently suppressed any discussion of retreat in his source text of Vegetius, the fleeing of the Dragon (the king's enemies) from the Panther (Christ) is made all the more condemnable.

1055 *princys wel appreved*. Vegetius specifically lists Xerxes, Darius, and Mithridates as examples of kings with extremely large armies. Persian emperor Xerxes I (r. 486–65 BCE) led an invasion of Greece that failed, in no small part, due to the large size of his army. Persian emperor Darius III (r. 336–30 BCE) lost his empire to Alexander the Great, who defeated him with a smaller army. After being defeated by Pompey, King Mithridates VI of Pontus (r. 120–63 BCE) faced rebellion trying to force a populace into armed service and committed suicide. Our poet, no doubt, suspects these examples will be less familiar to his audience.

1092 *affrayeth*. *MED affraien* (v.1), sense1b cites this line with the sense "harasses." We think it unlikely that the wind is harassing the flowers here, however. More likely, the winter wind is physically wearing them away, from *MED affraien* (v.2).

1098–1100 *First . . . . every welth*. D&A suggest that it is "by misunderstanding" (p. 135n1094) that our poet-translator omits Vegetius's notes on the need of the generals and officers to maintain the health of the ordinary soldiers in their army, instead replacing it with this passage on the health of the leadership being of primary importance. It is hardly a mistake, however. The late Middle Ages viewed the head of a body politic as being akin to the head of a body physic: the health of the head dictated the health of the body. For an introduction to the concept within the late fourteenth-century work of John Gower, for instance, see Peck, *Kingship and Common Profit*. As a political model, the association was notable within the Wars of the Roses, wherein each side could argue that the head of the opposition was unhealthy — physically or morally — and thus detrimental to the health of the realm.

1121–24 *The moneyles . . . . prince is lent*. The poet's advice here has no correspondence in Vegetius. Instead, his admonition against usury, his advice that re-payment should be made as soon as possible, and his assurance that money lent to the king is money that cannot be lost, all appear aimed at his contemporary audience. The late Lancastrian government held enormous amounts of debt from the Hundred Years War, a reluctance to raise taxes, and a stagnating economy. So well-known was the crown's despondency that it was a complaint of the rioters during Cade's Rebellion in 1450: "the king himself is so beset that he may not pay for his meat and drink, and he owes more than ever any King of England ought" (Ed. Myers, *English Historical Documents*, p. 267). As magnates pulled away to the Yorkist cause, the Lancastrian government was increasingly dependent on the good will and open purses of its supporters, who were increasingly in need of reassurances that they would be repaid.

1160–73 *And best bewar . . . . an hevy cas*. The poet goes considerably beyond the far more simple admonition in Vegetius that commoners, deceived by the false oaths of the enemy, can introduce strategic issues. It may be, as D&A suggest, that this is due to "the impression of sundry glaring acts of perjury in the course of the wars of the Roses" (p. 136n1160 ff.).

1170–73 *Judas . . . . an hevy case.* Judas's suicide by hanging after his betrayal of Jesus is related in Matthew 27:3–5. The poet's conclusion that this is a *hevy cas* (line 1173) can be read either to indicate the sorrowful nature of the allusion or to reconfirm the heaviness of the sinner's body weighed down by sins.

1202–08 *To falle . . . . undirtake.* The previous list of reasonable ways to occupy an idle army's time — military and physical training — parallels that given by Vegetius. Here, the poet adds a second list of rather more difficult tasks: reducing prominences, leveling broad fields, and tunneling through the earth.

1266–67 *trompe and clarioun / And pipe or horn.* On the uncertain identity of these various instruments, see the note to lines 782–88, above.

1279–92 *Semy-vocals . . . . declareth cleer.* On these instruments, see note to lines 782–88, above.

1296–97 *The sugnys mute . . . . not erre.* Stars, portcullises, and suns are among the many symbols that make an appearance in heraldry during the Wars of the Roses. Whether the poet has specific livery in mind beyond this general usage is not known.

1335–41 *Tak gidis . . . . hope and drede.* It was common for military leaders to seek out local guides while out campaigning, as detailed knowledge was not transmitted in any form beyond hearsay, with few exceptions. See, for example, Boardman, *Medieval Soldier*, p. 114.

1350–55 *Mynotaurys mase . . . . is sureste.* In Greek mythology, the Minotaur was a half-man, half-bull creature who lived in the Labyrinth, a maze so elaborate that those who went in could not find their way out before the Minotaur devoured them. The Minotaur was indeed an early Roman standard (Dušanić and Petković, "Five Standards"), and it was, as referenced here, symbolic of the need for plans of the generals to be kept as secret as the way through the beast's fabled maze.

1389 *helmettis.* Vegetius says nothing of helmets being given to those in charge of the baggage-train: instead, he emphasizes that officers were assigned the task of marshaling the baggage-train, utilizing a series of signals to coordinate its control in the event of an attack. It may be, as D&A assume, that *helmettis* here is meant to indicate these officers, "distinguished by the helmet from the camp-followers placed under their command" (p. 140n1389), but this is not certain.

1570 *wynne her shoon.* This echoes Froissart's famous claim that Edward III, upon hearing that his son, Edward of Woodstock (later called the Black Prince), was in danger at Crécy, told his men to "let the young man earn his spurs!" (Ed. Livingston and DeVries, *Battle of Crécy*, Item 81.540). This meant that the young Woodstock should be allowed to fight his way out of trouble himself, and so be deserving of spurs, which were heavily associated with knighthood and chivalry. When a knight was found to be treasonous his symbols were inverted or destroyed, including his spurs, which were chopped off.

1588–89 *That olde . . . . tary*. Instructive writings commonly assert that those in power, often the recipients of such volumes, should listen to the wise counsel of others, to better assert the value of the text in question. This is in a similar vein to the claims that good leaders read books, as mentioned in the note to lines 89–92, above.

1612 *if thin ooste is ace, and his is syis*. The poet's use of dice as an analogy cleverly relates both to the relative size of the armies — a single pip on a die (*ace*) being the smallest number possible and six pips being the largest — and to the likely outcome of battle between them. As noted in Chaucer's Prologue to the Man of Law's Tale, and likely alluded to in line 1619 here, higher numbers win in a typical game of chance (*CT* II[B1]124–25).

1633–41 *The chivalers . . . . in feithful diligence*. See note to lines 572–78, above.

1664–70 *Thus hardy hem . . . . a fyne corage*. Vegetius observes only that inexperienced men might think of flight on their first sight of battle. Perhaps with an eye toward reality, the poet accepts the likelihood that some of these men will indeed flee, and that they will need to be harshly rehabilitated back into the army.

1674–77 *The traditour Judas . . . . lyve and deye*. As in lines 1170–73 (see note above), the poet uses Judas's suicide as a symbol of the fate of traitors whose greed will defeat their own cause. While he is speaking here most directly to the general principles of instilling confusion in the enemy's ranks via *Rebellioun or ellis heresie* (line 1676), the poet no doubt recognized the specific occasions of such confusion in his own time: the Yorkist rebellion and, to a lesser extent, the Lollard heresy.

1678–84 *Oon thinge . . . . seyn the wise*. D&A suggest that this stanza "interrupts the flow of the English poet's thoughts" on "the effects of dissension as witnessed by himself in contemporary history" (p. 149 stanza 1678–1684). This is mistaken, likely due to D&A gravely misreading the source text in Vegetius. It is true that dissension was the topic of the preceding stanza of the poet's translation (lines 1671–77), and that his contemporary reflections upon dissension will be the topic of the next stanza (lines 1685–91), but these two thoughts are properly bridged by this intervening stanza. Continuing his close translation of Vegetius, the overall goal of the entirety of his book is recalled: that what has been done in the past can be recreated in the present. Contrary to D&A, for Vegetius (and thus for our poet) these comments are not pertinent to only "the task of reducing an undisciplined soldiery to military discipline, this being the subject-matter of the chapter"; they are, rather, foundational to the reformative goals of the entirety of the work at hand. This stanza, then, is no interruption of the "poet's thoughts," but it is instead a continuation of the translation of Vegetius's words, which serves to introduce the topic of the following stanza.

1680 *As hath be doon, it may be doon ayeyn.* In the course of translating Vegetius's principle of learning from the past, the poet is moved to paraphrase Ecclesiastes 1:9: "What is it that hath been? The same thing that shall be. What is it that hath been done? The same that shall be done."

1682–83 *In wynter colde, in somer dayis fayre / Is good to se.* D&A, missing the poet's analogy, posit the meaning "In winter, it is a good thing to experience cold weather, in summer — fine weather" (p. 149n1683). The lines instead refer to a human response to the passage of time, as in his example of the castle that falls into ruin from disuse and must be fixed (line 1681). Here, one should see in winter's chill the inevitable fair sun of summer that will come again. More than that, one should prepare for that summer, since actions taken in good months will bring about survival through the bad months. So it is with war and peace, as he goes on to explain.

1685–91 *In Engelond . . . . revyve and rise.* The reflections Vegetius made upon his past and present are swapped out for the poet's own. The idea that the English people in the 60 years prior to the First Battle of St. Alban's on May 22, 1455 had seen no war is hyperbole that neglects bloody engagements at Bryn Glas (1402), Shrewsbury (1403), Bramham Moor (1408), Agincourt (1415), and Orléans (1428), to name only a few. At best, a scribal error has made XL into LX, followed in C and R (leaf missing in A), and the poet intended to say that English lands had not seen major conflicts in roughly 40 years.

1687 *oon bataile aftir the blasing sterre.* Any of several comets from the period, including Halley's comet in June 1456, could be meant. Regardless, D&A note that "the author's sweeping assertion evidently disregarding all minor skirmishes, this . . . points to the Yorkist defeat" at Ludford Bridge on October 12, 1459 (pp. xx–xi), which is also referenced at lines 985–1026.

1688 *hem that whirleth as the fane.* That is, the Yorkists who, in the poet's view, continually promise peace — as they had at the Loveday of 1458 — only to "turn with the wind" and return to rebellion and conflict.

1690–91 See note to line 431, above.

1692 *the wepon bodeth peax.* Proverbial; see Whiting W152.

1706–12 *Here is . . . . us recommende.* This stanza is meant to mimic the commonly employed pre-battle sermons. One of the more famous examples of these is that given before the Battle of Agincourt (1415), which is found in the contemporary *Gesta Henrici Quinti* (Ed. and trans. F. Taylor and Roskell, pp. 78–79, 84–87). In these writings English piety is typically contrasted with the enemy's overconfidence and revelry the night before battle, often to make a direct allusion to the first-century-BCE Roman writer Gaius Sallustius Crispus's widely read *De bello iugurthino* (trans. Batstone, 98.6–99.1).

1709 *make us of this grete ernest a game.* An interesting inversion on Chaucer's warning that "men shal nat maken ernest of game" in the Prologue to the Miller's Tale (*CT* I[A]3186).

1734–40 *It is . . . . shal prevaile.* D&A here propose that the poet has given "exactly the contrary" advice of Vegetius, who suggests that if the most experienced fighters are afraid the general should postpone the fight: "This divergence reveals a fundamental difference in the attitude of the two writers. With Veg., the clear-headed, unemotional expert, it is a question of trusting the *experience* of soldiers, with the contemporary of feudalism, whose head is crammed with untechnical notions of chivalry and piety, it is a question of trusting the *courage* of 'knights'" (p. 151n1738). This reading is highly suspect. One, Vegetius is hardly as rationally objective as they claim. Two, the notion of "experience" is equally (if not greater) as "untechnical" as "chivalry and piety." Three, there is no "divergence" between the two writers, as their advice is the same: a leader should understand the *fere or confidence* (line 1736) of the men who are to fight. Veteran men who are fearful of the fight, as Vegetius recommends, ought not to be trusted to perform well in it. Likewise untrustworthy are inexperienced men who are over-boastful about the fight.

1748–49 *The first sight . . . . in fight.* Once again, D&A wrongly condemn the poet for differing "from Veg. in a characteristic way. For Veg. says that all men experience terror when a battle approaches. The English priest [i.e., the poet], and admirer of feudal chivalry, restricts this statement to those '*that neuer were in fight*'" (p. 151n1748 f.). However, the poet says only that the inexperienced men are likely to be the *ferdfullest*: for these to be the "most fearful" requires that other men, in accordance with Vegetius, be fearful, too.

1790–91 *This ege in dayis olde a principaunt / Of wurthi men, as princys, had his name.* The poet somewhat mis-renders Vegetius here, who speaks of the first line being made up of the most experienced men, who were thereby called the *principes*.

1803 *DC*. D&A follow MS in printing *CC* (wrongly implying all MSS follow this reading), but C, R, presumably A (R is copied from it), and Vegetius (as well as simple mathematics) shows that this is an early scribal error for *DC*, rendering the number of infantrymen as 1,666 instead of 1,266. The poet gets the number correct in line 1870.

1821 *Until thei yeve her coors to seyntewary*. D&A inexplicably suggest this to mean "surrender," which is noted as being "not in Veg." (p. 154n1821). *MED seintuarie*, sense 3b, follows suit, citing this line. This would require a rather cumbersome reading in which the enemy will fight "until they give their bodies into the claim of immunity from arrest," with the implicit understanding that this means immunity from being killed. To the contrary, *MED seintuarie*, sense 4a allows for a far more straight-forward sense that the enemy will fight to the death, "until they give their bodies to sanctuary (i.e., burial ground)."

1838 *on hem reyse a smoke*. Proverbial; see Whiting S413.

1839 *foomen*. It is possible, perhaps likely, that the scribe has here miscopied *foomen* for *footmen*: men on horse chase down the men on foot. The line is also readable, however, as it stands: the horsemen chase down the men who are their foes (see *MED fo-man*, n.). Given the consistency of the word across the surviving manuscripts, we have opted to let *foomen* stand.

1846–52 *The fifthe cours . . . . VIII or IXne.* This fascinating passage on ranged artillery moves from the technical terms for specific Roman ballistic weapons to those for a range of contemporary artillery pieces. In doing so, it neatly summarizes the movement not just from ancient to medieval warfare, but also the remarkable impact of gunpowder weaponry. The *carrobaliste* (Latin: *carroballista*; line 1846) is usually understood to have been a cart-mounted ballista in the Roman army, although some scholars think it could mean a ballista that is transported by (but not shot from) a cart. The *manubalistys* (Latin: *manuballistae*; line 1847) were hand-carried ballistae, not unlike crossbows. The poet's fundibulary refers to Roman staff-slingers (Latin: *fundibulatores*; line 1847), who used a staff sling (Latin: *fustibalus*) to hurl stones at the enemy. The *funditours* (line 1848), by contrast, used a more basic hand sling to do so. Much of this Roman war-machinery is in the poet's opinion now *unwiste* (line 1848) with the advent of gunpower weaponry. The largest of this new artillery is the bombard (line 1849), which could fire stones of enormous sizes (5–10 tons was common). The serpentine (line 1850) was smaller (firing 500 lb. balls) and often found in naval usage. The even smaller veuglaire (*fowler*; line 1851) had a wide range of sizes, and it seems to have been differentiated from the crapadeau (line 1851) by the ratio of its length to its bore. The coulovrine (line 1851) would have been smaller still, and meant to be carried by hand. The poet's *covey* (line 1851) is unidentified, but descriptions of the rest of the guns mentioned here can be found in R. Smith and DeVries's *Artillery of the Dukes of Burgundy*, pp. 204–36. Interestingly, this appears to be the only reference to a crapaudeau in English, which may well point to the poet's familiarity with Burgundian and other continental artilleries, a familiarity he no doubt would have gained from his time in Calais. These artillery pieces appear once again in lines 2854–67.

1853–54 *that hadde as yet no sheeld / As bachelers.* The shield was the traditional location to display one's coat of arms. The poet here further explains that the knight bachelor was a knight who had yet to earn the right to fight under (or behind) his own banner by stating that he *hadde as yet no sheelde*. See also note to lines 292–93, above.

1859 *In theym, as in the thridde, al to repayre is.* The poet's sense is unclear, but it seems to be some attempt to render an etymology for the Latin term *triarii*.

1887 *Untaught in Doctrinal or in Grecisme.* The poet refers here to two widely read Latin grammars of the Middle Ages: Alexander of Villedieu's *Doctrinale puerum* (written ca. 1200) and Eberhard of Béthune's *Graecismus* (ca. 1212).

1926–34 *if the boorys hed . . . . as beforn.* This remarkable image, not found in Vegetius, likens an attacking wedge of men to a boar's head, just as the boar uses its pointed snout and tusks to dig around the tree-roots in a forest; the wedge uses its pointed formation and sharpened weapons to charge into a line of armed men. The proper leader will respond, the poet says, by reinforcing the line and thus cutting off the boar's snout — the point of the attacking wedge — at its root.

1935 *yawe.* A "saw" is the formation that a reserve unit takes when part of the main line grows disorganized: the reserve unit rushes to the front and forms a temporary advanced line against the enemy, behind which the main line can reorganize.

1986–92 *A clamour . . . . not incredibil.* There are countless descriptions of combatants yelling their battle cries, often the names of their leaders or their cause, right before clashing with their enemies. These reinforce the importance of identity to medieval warriors, also commonly expressed through heraldry.

1995–96 *With myghti countynaunce, that is the mon / That mornynge is to have a ful fayr day.* Vegetius emphasizes the benefits of being the first to draw up lines and attack, but this further comparison, original to the poet, shows the translator's distinct flair for welding his personal knowledge to his task. Acting first in battle, he says, is like the moon that presages a fair day. This simile is, first, nautical in its background. A saying among sailors even today is that "A ring around the sun or moon means rain is coming soon" — that is, a halo around sun or moon is caused by light refracting through cirrostratus clouds, a strong indication that storm clouds are on the way — and, vice-versa, a clear moon at morning indicates a clear day. In addition, the simile is theological. God, according to the prophet Isaiah, will bring ultimate victory to the people of Zion: "there shall be upon every high mountain, and upon every elevated hill, rivers of running waters in the day of the slaughter of many, when the tower shall fall. And the light of the moon shall be as the light of the sun" (Isaias 30:25–26).

It is possible, too, that the poet has introduced a further level of word-play here, as the *mon* could refer to a moan and the *mornynge* could refer to mourning. Thus, attacking first induces the moan that causes mourning in the enemy host, a fair day for the attackers!

1999 *palme of victory.* Though Vegetius does not use the image here, the palm of victory is a classical symbol: victorious Roman leaders and fighters could be presented with them. For the poet, the image carries additional weight: during Jesus's entry into Jerusalem, as described in John 12:13, believers "took branches of palm trees and went forth to meet him," giving rise to the Christian tradition of Palm Sunday. In Apocalypse 7:9, these symbolisms are fused: those marked for salvation stand before the throne of God with "palms in their hands."

2007–48 *Here angelike valiaunce . . . . in eyther horn.* Perhaps inspired by the account of counter-actions in Vegetius, the poet digresses into a discussion that appears to juxtapose the celestial glory of King Henry VI and his army — recalling the poem's opening stanzas — with the cursed nature of his Yorkist enemies.

2017 *A legioun attaynte.* During the 1459 Parliament, which was held in Coventry between November 20 and December 20, bills of attainder were passed against twenty-nine leading Yorkists.

2019–20 *Her lord . . . . his side.* Lucifer has long been seen as a figure of pride in Christian thought, ultimately building out of Isaias 14:13–14.

2088 *bak and breeste.* By careful word choice, the poet manages to create an image that perhaps works in both the macrocosm and microcosm of combat. The direct meaning is that as the opposing formation is caught between battle-lines, each man in it suffers attacks to both his breast (*MED brest* [n.1], sense 2a) and back (*MED bak* [n.], sense 1a). At the same time, the larger formation itself might be said to be attacked from both front (*MED brest* [n.1], sense 6a) and rear (*MED bak* [n.], sense 9a).

2128 *But fro behinde aboute is beste it brynge.* D&A pronounce that the line "is not clear" and suggest that it means that the commander, having freed one wing of his army by fixing his position upon flanking terrain, ought to send the free wing to attack the enemy from the rear (p. 162n2128). While such a maneuver would be of clear advantage, it is not what the poet means to convey here. To the contrary, the poet is instead observing that the commander ought to move the free wing — the referent of *it* — behind his own established lines rather than moving it across their face. Following this advice would both minimize the enemy's awareness of it and minimize any potential confusion in the commander's own lines.

2140–60 *Yet heer . . . . of victory resoun.* Vegetius writes about the advantages and dangers of ambush that exist whether an army is on the attack or in retreat. The poet, as D&A observe, "resolutely refuses to admit the possibility of 'his' side being forced to retreat at all," and so greatly condenses Vegetius's advice (pp. 162–63n2151–3) and reflects some extreme chivalric ideals in the later Middle Ages. See also note to lines 275–77, above.

2165–66 *Al suffisaunce / Hath he that is content.* Proverbial; see Whiting S867.

2166–67 *Al may be born / Save wele.* The poet's advice here is difficult to discern. It is possible that he is hinting at something like the oft-repeated wisdom of Ecclesiastes 5:5 (see also Job 1:21, Psalm 49:17, 1 Timothy 6:7) that riches cannot be carried into the next life. Alternatively, the poet is commenting on the need to focus on winning the encounter before engaging in the division of loot or the taking of prisoners for profit.

2167 *scorned is that useth scorn.* Proverbial; see Whiting S92.

2168–69 *Thi disavaile availe is to thi foo, / His hurt availeth thee.* Though the poet here has returned to his source, the maxim is essentially proverbial; see Whiting H653.

2177 *Lerne of thi self disgised.* D&A suggest that the sense is to "learn of your own will, keeping your intentions secret" (p. 163n2177). Another possibility is that the sense is to "learn from questioning yourself."

PART IV: URBAN AND NAVAL WARFARE

2226 *The morthereer to bringe undir the cheyne.* D&A posit a Lancastrian partisanship to the line (p. 165n2226), and there is no doubt that the poet elsewhere makes clear his opinion that Yorkists are unlawful murderers. Nevertheless, there is no grammatical sense in which this line need only apply to the enemy. To the contrary, the theologically trained poet would recognize that killing for the state, even when sanctioned by the Church, is nevertheless murder. A manual of warfare, such as *Of Knyghthode and Bataile*, is at its heart an attempt to bring order (*the cheyne*) to what would otherwise be the chaos of murderers on the rampage. Indeed, the Yorkist text, in C and R (leaf missing in A), make no changes to this line at all.

2231 *Brutis Albion*. Here the writer alludes to the legendary foundation of Britain by Brutus of Troy, descendant of Aeneas, the Trojan hero of Vergil's *Aeneid*. The island of Britain was originally named Albion, but after Brutus arrived he renamed it Britain. Brutus's kingdom, after his death, was split into England, Scotland, and Wales between his three sons. England was given to his oldest son, Locrin, who the English kings claimed to descend from, in one way or another, and thus claim overlordship of all of Britain in the later Middle Ages. The English origin story was recounted in a number of texts, but most famously in the *Prose Brut* chronicle, which initially ended at 1272 and explicitly connected the legendary stories with contemporary rulers. The *Brut* was originally written in Anglo-Norman French and received several continuations and translations into Middle English (with further continuations) and Latin. This English origin story served to trace England's authority to rule back to the Trojans, a strategy similarly employed by other medieval kingdoms. See Matheson, *Development of a Middle English Chronicle*, and, more recently, Marvin, *Construction of Vernacular History*.

2233 *To Normandie and Fraunce it is t'assende*. Though one might be tempted to read into the verb here some Continental bias in the idea of "ascending" to European shores, it is more likely that the poet means one must ascend a ship (i.e., board it) in order to reach the Continent.

2273 *the inner wal . . . . then withoute*. Rather than two separate walls, the poet is referring here to the inner and outer faces of a single construction that is more technically termed a terraplein.

2315–16 *Glew, tar . . . . brenne engyne*. Despite the use of gunpowder weapons, wooden siege engines were still common in this period, so the preparation of incendiaries is sound advice. These wooden engines were sometimes made off-site and kept in storage, such as at the Tower of London, and brought to sieges, or, in the case of simpler engines like rams, were made on-site from local materials.

2319 *shelde and spere, hundirdys VIII or IX*. The specificity of the numbers involved — which are not in Vegetius — are only meant to indicate the high quantities at hand (and to meet rhyme).

2333 *pulle over their topsail*. The use of nautical imagery here, which is original to the translator, is one more indicator of his possible personal experiences with the sea. See also the note to lines 1995–96, above.

2337 *crosbowyng . . . bowe of brake*. The poet appears to differentiate here between crossbows whose strings are pulled back physically by a lever and those whose strings are pulled back mechanically by the use of a winch or *brake*.

2341–42 See note to lines 2959–65, below.

2357–59 *A see citee . . . . litil season*. This method of harvesting salt is still practiced today.

2371 *tortoys or the snayl.* If the poet has any real-world differentiation between these terms, it is not clear. They appear instead to be redundant terms for a Latin *testudo*, meaning "tortoise": a protective shell beneath which a ram or siege-hook could be utilized. These weapons would poke in and out from beneath the shell, as the head of the animal pokes in and out from its shell.

2372 *sekel or the sithe.* Siege-hooks were used to chip out the stones at the base of a wall.

vyneyerde. Protective sheds could be used across the field in order to protect siege operations. As described in lines 2427–40, a number of these sheds could be lined up to cover a long trench, as needed to support mining operations, and would thus appear as vines across the ground.

2381–84 *It . . . . moost avavile.* The "horns" described here by the poet are of unknown purpose; certainly they do not appear in the text of Vegetius.

2385 *tortoys, snail, or testude.* See note to line 2371, above.

2392 *Tweyne hornys if it have, it is a snaile.* What the poet imagines as the "snail" is unclear; this construction has no corollary in Vegetius.

2399–2426 *And when . . . . perile and peyne.* Having just defined the construction of ramming weapons, the poet has chosen to turn immediately to Vegetius's discussion of defense against them. This requires a break in his general practice of adapting his source text in sequential order.

2410 *wulf.* As the text describes it, this "wolf" is a device intended to be lowered down from the wall in order to pinch a ram in its teeth and prevent it from striking the wall itself.

2427 *vyneyerde.* See note to line 2372, above.

2436–40 *And tre pilers . . . . be myscheved.* Mining was an effective, if time-consuming, method of bringing down the enemy's wall. Unfortunately, clear evidence of mining has only survived in a few cases. One of the most famous examples is the incomplete mine, as well as its counter mine, made under St. Andrews Castle in Scotland in 1546.

2488 *molliols, also fallayrys.* As Vegetius describes them, these fire-darts (Latin: *malleoli*) and fire-spears (Latin: *falarica*) appear to be just what they are described as: bolt-like darts or spears that are covered in flammable material, set ablaze, and cast against the enemy by either hand or engine.

2558–59 *And sodenly . . . . therate.* The poet here omits further details of besieging forces pouring through the mines and killing people in their own homes, which D&A claim is because "this was more than the chivalrous temper of the Englishman could stand" (pp. 173–74n2555). Although the poet makes some clear alterations to his source material to support his own perspective, it is clear that all combatants in English armies, including knights, killed civilians in war, especially in sieges. There seems to have been an unspoken custom of the "right of storm," in which if a city or castle had not surrendered but was taken by force, then all property and people therein were forfeit to the conquerors and could be treated however they desired (Keen, *Laws of War*, pp. 119–33, and Strickland, *War and Chivalry*, pp. 222–24). However, details of such actions after the fact clearly have an immoral tone to them that requires further qualification to make them appear just, and writers frequently express anxieties over portraying such actions by their own people (T. Smith, "National Identity," pp. 79–110, 141–48, 183–87). It should be noted that the English were not as chivalrous or careful in their conduct of war as D&A suggest in their romanticized claim.

2565–66 *they wil . . . . debate is.* The poet portraying the conquest of cities and castles as less violent by offering the enemy a chance to leave is not unique to English writers, despite D&A's suggestion that "it is characteristic of the chivalrous bravery of the Englishman" (p. 174n2565 f.). Instead, it more expresses the common anxieties over the ethics of the "right of storm," but also echoes the argument of Veg. 3.21, in which cutting off all hope of escape in the enemy is ill-advised, as surrounded men fight harder.

2595–2601 *Now se . . . . to calle is.* The placement of guard dogs (and other noisy, excitable animals) on or near walls is mentioned in many Roman sources, which Vegetius clearly drew upon. Medieval chroniclers also describe such defensive measures, although it is unclear whether these were merely classical allusions or reflected contemporary practice.

2609–2972 *The craft . . . . other keste.* Vegetius's discussion of ships and naval warfare found increasing interest in the fourteenth and fifteenth centuries, especially in how they advocated for standing navies of specially constructed warships. In several manuscripts this section of the text is separated into a fifth book (Allmand, *The "De Re Militari" of Vegetius*, pp. 44–46). Here, the poet greatly and imaginatively expands on his source, emphasizing the importance of naval warfare despite Vegetius's suggestion that the less said of it the better since there were few sea-battles in his time, as by then all of Rome's major Mediterranean enemies had been subdued. Things are far different for the fifteenth-century poet: his world saw far more naval engagements, and his personal history in Calais would have made him particularly aware of the importance of naval power even if he had no personal service at sea in his background.

2612–15 *Now go . . . . gret nede.* Naval battles were understood to be considerably different from land battles, with often bleak consequences for the defeated (see, for example, lines 2896–2902).

2687–99 *Tethis . . . . be stille.* Vegetius does not personify the sea as Tethys, the Titan sister-wife of Oceanus in Greek mythology, nor does he pursue anything like the poet's metaphor of the (masculine) winds assaulting the sea against her will. Likewise, the further description of a disturbed Oceanus leaping against the town of Calais (line 2697) is original to the poet, who claims himself to be from Calais and would surely know well the vagaries of weather in the English Channel. The poet may also be allegorizing the contemporary struggles between Yorkist and Lancastrian forces within the disputed waters at the time of his writing. Wakelin suggests that these descriptions might be an allusion to Book I of Vergil's *Aeneid* (*Humanism, Reading, and English Literature*, pp. 87–88).

2718 *the berth of Arcture al is even. Arcture* could refer either to the star Arcturus, in the constellation Boötes, or to Arthur's Plow, another name for the constellation Ursa Major (i.e., the Big Dipper) (see *MED Artur,* [n.]), though the former is surely meant here: in September, Arcturus — the fourth brightest star in the night sky — rises close to *even* with the horizon, its place of *berth.*

2768–69 *In wynde . . . . begile.* The poet (following Vegetius) refers here to Virgil's *Georgics,* trans. Fairclough, 1.351–463, in which appears a wide-ranging list of terrestrial signs regarding the weather.

2791 *the maister marynere, the governour.* Allmand argues that because sea storms were typically used to symbolize political turbulence, the use of *maister marynere* here paired with *governour* was meant to comment upon the leader's ability to govern: "he was a man worthy to be entrusted with the ship (of state) and its crew (the people), a man in whom to have confidence to reach land — and safe harbour" (*The "De Re Militari" of Vegetius,* p. 192), while Wakelin reads this as indicating the leader's obligation to his land and people (*Humanism, Reading, and English Literature,* pp. 89–90). The poet's proximity to the sea, in Calais, suggests he surely found this imagery especially powerful.

2833–2909 *But on . . . . your renoun.* The poet deftly maneuvers from his translation of Vegetius into a dramatized description of the naval battle of Massilia (49 BCE). The poet's visceral description of the course of this engagement has many echoes in the naval warfare described in the first-century Roman poet Marcus Annaeus Lucanus's widely read *De bello civili.* Though what the poet provides thus appears to be an imaginary engagement, there is no question that it is intended to reflect a contemporary struggle at sea between his Lancastrian allies and the Yorkist rebels. Indeed, D&A rightly suggest this is meant to function as a kind of prophetic vision of what is to come in the war (p. xxi).

2854–67 *The canonys . . . . and the manubaliste.* D&A (p. lxx) believe that the poet's listing of these different terms was meant to demonstrate his familiarity with gunpowder weapons, regardless of how familiar he actually might have been with their actual uses in warfare. It should be noted, however, that he correctly describes their use. Likewise, he initially presents both the ancient and medieval weapons in apparent order of weight earlier in the poem. See the note to lines 1846–52, above, which also identifies the pieces.

2857 *goth the serpentyne aftir his mete.* The poet cleverly doubles the meaning of the line by playing with the bestial nature of the name of the artillery piece: the serpentine gun goes after its target, just as the serpentine creature goes after its meal. Similar metaphors continue from line 2861.

2866 *onagir.* The onager was a small, Roman-era catapult.

2868 *catafract.* This appears, as noted by D&A (p. 183n2868), to be "a mistake of the poet" deriving from his misunderstanding of Latin *catafracti* (which appears in Veg. 4.44), meaning "armored."

2875–81 *Yet on . . . . on hem ride.* This description of the superior English archery and the fear it instills in the enemy is not uncommon. The English chronicler, Geoffrey le Baker (d. ca. 1360), provides strikingly similar imagery in his narrative of the 1350 naval Battle of Winchelsea (trans. T. Smith, "Battle of Winchelsea").

2896 *fisshes for to fede.* This grizzly image of fish eating those who have been recently killed or drowned in naval battle is sometimes mentioned in medieval sources, such as the description of the naval battle of Winchelsea (1350) by the English poet Laurence Minot (fl. 1333–52): "Fele fissches thai fede for all thaire grete fare; / it was in the waniand [waning of the moon, i.e., an unhappy hour] that thai come thare" (ed. Osberg, 10.5–6). Wakelin reads our poet's use of this imagery as reflecting a Lancastrian "glee" over their defeated enemies (*Humanism, Reading, and English Literature*, p. 88).

2899 *summe seke hemself they wote ner where.* In glossing this line we have followed D&A (p. 184) in assuming that this is a continuation of the thought about fearful men in line 2897: the poet is now describing those panicked individuals who are desperate to find safe places to hide but do not know where to find them. However, the syntax seems to allow for at least two alternative readings that delve even more deeply into the horrifying realities of war. The first would be that we have a reference to mental breakdowns among the crew. In the trauma of the combat, the men have lost their situational awareness of reality: they seek themselves as they no longer know where they are. The second alternative would be that the poet is describing a grim but horrible reality of gunpowder combat: dying men are seeking those parts of their bodies that have been physically lost.

2908–09 *O Silver Bere . . . . your renown.* The poet returns again to his heraldic allegory to refer to the rebel enemy of the War of the Roses (see notes to lines 992–1026, above): the Bear is Richard, earl of Warwick; the Lion is Mowbray, duke of Norfolk; and the Eagle is Richard, duke of York.

2916 *the knotte.* Knots have long symbolized the mystery of the Church, whether in artistic representations like the trefoil knot or in literary references such as this ca. 1465 attack on Lollardy: "Lo, he þat can be cristes clerc, / And knowe þe knottes of his crede, / Now may se a wonder werke" (Ed. Robbins, "Defend Us from All Lollardry," lines 1–3).

2941 *polissed to bace and make it herre.* The ability to raise and lower this ramming beam is not mentioned in the source text, Vegetius.

2947 *is hard to wite.* The mariners are clearly using axes to take down the ship's rigging. What is less clear is whether the poet is remarking on it being difficult to understand what the mariners are doing or he is commenting on how hard it is to accept it. The latter possibility could point to a personal nautical background.

2959–65 *Theim lever . . . . bodily beautee.* The poet's immediate source for this anecdote is Vegetius, but there are multiple similar stories across classical sources. As Nall points out, the poet adds "worldy good or bodily beautee" to his source to emphasize the sort of sacrifice that was expected of everybody in England, not just its men, to succeed in war ("Perceptions of Financial Mismanagement," p. 133; Reading and War, pp. 65–66).

2968 *Oon more of hem then X manubalistis.* D&A make multiple mistakes in suggesting that this is "a clumsy way of saying 'eleven'" and that manubalistis is used "to denote the men manipulating the *carroballista*" (p. 185n2968). The poet is consistent in using *manubalistis* to refer to a weapon of similar design but smaller scale to the *carroballista*. This same concept of relative size is here described: the larger weapon takes more than ten times as much sinew to fire.

2982 *That may not Joon! Whi so?* While it could be that the word *Joon* (if that reading is correct; see the Introduction, pp. 16–18) is a reference to the name John and thus an indicator of the author's identity, in the Middle Ages the name John was commonly used as a generic, familiar name for a priest (which the author is known to have been). See, for instance, the Host calling to the Nun's Priest in Chaucer's Nun's Priest's Prologue: "Com neer, thou preest, com hyder, thou sir John!" (*CT* VII[B2]2810).

2986 *world withouten ende.* The Gloria Patri, often termed the Minor Doxology, is one of the oldest and most universal doxologies in Christianity. It concludes in Latin with the phrase "in saecula saeculorum," which since the release of *The Book of Common Prayer* in 1541 has been popularly translated as "world without end."

2988 *until that ayer amende.* For the poet, there were two problematic heirs in Calais. One would be York's son, Edward, earl of March, who had fled there alongside another rebel heir: the earl of Warwick, who was in nominal control of the city.

2995 *laborers.* Here the poet seems to refer back to lines 180–200, in which he argues that country folk tend to make the best fighters.

3013–14 *VII / Weyis.* These are the directions given for the setting of pitched battles in lines 2046–2132.

3022 *Go, litil book.* In humbly sending his "little" book into the world (see also line 2987), the poet is taking part in a fairly standard trope that appears to mimic the "Go little book" envoi of Chaucer's *TC* (5.1786).

TEXTUAL NOTES

In accordance with METS policy, we have regularized *the* to *thee* when it refers to the second-person pronoun in the following lines: 53, 1104, 1172, 1614, 1650, 1759, 1999, 2080, 2169, and 3023. Likewise, we have regularized *of* to *off* in the following lines: 113, 189, 949, 1410, 1541, 1932, 1948, 2004, 2057, 2127, 2369, and 2471.

Abbreviations: **A**: Oxford, Bodleian Library, MS Ashmole 45 (Part 2), fols. 1r–7v, 18r–23v, 41r–43v, 46r–53v; **C**: London, British Library, Cotton MS Titus A.xxiii, fols. 2r–53v; **D&A**: *Knyghthode and Bataile*, ed. Dyboski and Arend; **MS**: Cambridge, Pembroke College, MS 243, fols. 1r–55v [base manuscript]; **R**: London, College of Arms, MS R.25, fols. 24r–62v.

Due to errors in the prior edition of the text (D&A), and the more recent discovery of a fourth copy of the text (R), we here record the substantive differences between all four known copies, omitting only minor orthographic differences. Lines that appear to have been changed due to Yorkist influence on the manuscript tradition are highlighted in **bold** font.

1–88 *Salve . . . . Amen.* Omitted in C, A (possibly copied from C), R (possibly copied from A).

6 *chare.* So MS. D&A: *chaire.*

7 *comyn.* So D&A. MS: *comyn . . . with . . .* illegible even under UV light.

10 *anende.* D&A: *an ende.* MS: *an ende.* See Explanatory Note to lines 9–15.

12 *on.* So D&A. MS: *or.*

25 *gratias.* So MS. D&A: *gracias.*

53 *preste.* MS: fully legible only under UV light. D&A: *preste.*

55–56 *Lo . . . . the gise.* Written in reverse order in MS, corrected with marginal notation.

55 *Anoon.* So MS. D&A: *anon.*

57 *servyse.* D&A: *seruyse.* MS: *srvyse*; the lack of proper abbreviation is unnoted by D&A.

77 *theryn doeth.* MS: fully legible only under UV light. D&A: *Ther, yeueth.*

91 *And have.* C damaged.

exercise. So MS, C, A. R: ~~*exdce*~~(?) *exercise.*

92 *Pleasaunce.* C damaged.

emperour. So MS. C, A, R: *emperours.*

93 *O.* C: damaged.

100 *out.* So MS, C, A. R omits.

the. So MS, C, A. R: *thy.*

103 *Maria.* So C, A, R. MS: fully legible only under UV light. D&A: *Maria.*

104 *myst.* So C, A, R, D&A. MS: fully legible only under UV light.

106 *of.* So MS, C, A. R: *and.*

ye. So MS, C, A. R: *the.*

109 *Remembering.* A damaged.

116 *Help heer to make of werre.* So MS. C: *Help here to make of werre.* A: *Hel here to make of werre.* R: *here of warre to make,* with an unknown word erased from the beginning.

118 *ignoraunce.* C damaged.

119 *propre speche.* C damaged.

120 *pleasaunce.* C damaged.

121 *Harry.* So MS, D&A. C, A, R: *Edward.*

Fraunce. C damaged.

125 *His Modir, and His sayntis.* So MS, C. A: *His Modir and His saintis.* R: erased.

136 *maundement.* So C, A, R. MS: fully legible only under UV light. D&A: *mandement.*

140 *and*$_2$. So MS, A, R. C: *in.*

143 *ooste.* MS: fully legible only under UV light. C, A, R, D&A: *oste.*

145 *Emanuel.* A damaged.

146 *Hath ostis.* A damaged.

147 *That noon of.* A damaged.

wote. So MS. C, A: *wot.* D&A: *woote.*

148 *Their.* A damaged.

150 *truely.* So MS. C, A: *truly.* R omits.

ful fer. So MS, C, A. R: *sulfer.*

155 *vertu.* So C, A, R. MS, D&A: *vertue.*

160 *hosteyinge.* So C, A (*h* scraped). MS: fully legible only under UV light. R: *hasteying.* D&A: *hosteyinȝe.*

172 *see.* So C, A, R. MS: fully legible only under UV light. D&A: *see.*

175 *blood.* A damaged.

176 *thei drede it.* So MS. C, A: *her drede is.* R: *her dryd is.*

178 *blood.* So MS, A. C: inserted above the line. R: *blod.*

181 *atwix.* So MS, C, A. R: *~~bx~~ atuex.*

182 *over hote.* So MS. C: *over hot.* A: *over ~~hote~~ hot.* R: *hoot.*

184 *feeldys.* So MS. C, A: *feeldis.* R: *in feldis.*

184–85 *In colde . . . . beth used.* Written in reverse order in A, corrected with marginal notation.

186 *chere.* So MS, D&A. C, A, R: *chese.*

188 *that.* So MS, C, A. R omits.

189 *is.* So MS, C, A. R omits.

190 *and*$_2$. So MS, C, A. R: *of.*

211 *That.* So MS, C, A. R: *Then.*

here yougthe in. So MS. C: *her youthe in.* A: *her youghe in.* R: *in yough here.*

212 *yonge may.* So C, A. MS: *yonge ~~man~~ may.* R: *yong may.*

214 *in eny.* So MS (*in* inserted above the line), C. A: *in any.* R: *in no.*

219 *there.* So MS. C, A: *ther.* R: *therein.*

220 *is.* So MS, C, A. R: *it is.*

222 *sixe.* So MS, A, R. C: *~~sy~~ sixe.*

227 *Lo.* So MS, C, A. R: *To.*

228 *in vigour had no pere.* So MS, C, A. R: *se lese in vigour had pere.*

229 *to*$_1$. So MS, C, A. R omits.

232 *hondys.* So MS. C, A, D&A: *hondis.* R: *houndes*(?).

233 *is.* So MS, A, R. C: *~~to~~ is.*

234 *in.* So MS, C, A. R: *Virgilius in.*

243 *So.* So MS, C, A. R: *~~So~~ So.*

253 *se.* So MS, C, A. R omits.

256 *forlafte.* So C, A, D&A. MS, R: *forlaste.*

is. A damaged.

257 *and.* So MS, C, A. R omits.

258 *boor.* So MS. R: *wild boor* (*wild* added from interlinear gloss in A: *wilde*; or C: *wild*).

263 *have.* So MS, C, A. R: *hem.*

crafte. So MS, C, A. R omits.

grete. So MS, C, A. R: ~~*gete*~~.

275 *hem*. So MS, C, A. R omits.

281 *his*. So MS, C, A. R: *is*.

282 *hym bere*. So MS, C, A. R: *bere hym*.

286 *that have habilité to werre*. So MS, C, A. R: *abilite of werre*.

296 *he*. So MS, C, A. R: *it*.

298 *preef*. So MS, C, A. R: *prief*. D&A: *pref*.

303 *Is*. So MS, C, A, R. D&A: *In*.

305 *whilom was*. So MS, C, A. R: *was sum tym*.

307 *informal*. So MS. C, R: *informall*. A: *infourmall*.

309 *th'istories*. So MS, C. A: *th'ystories*. R: *in storris*.

312 *or*$_1$. So MS, C, R. A: ~~*of*~~ with *or* inserted above the line.

315 *amende*. So MS, C, A. R: *to amend*.

323 *seyn*. So MS, C, A. R: *her seyn*.

324 *jumpe*. So MS, C, A. R omits.

332 *it*. So MS, C, A. R omits.

sture. So MS, C, A. R: ~~*sure*~~ *seure*.

338 *blood*. So MS, C, A. R: *of blood*.

339 *and*. So MS, C, A. R omits.

343 *eson*. A: rebound too tightly to read.

344 *rayn*. So MS. C, A: *reyn*. R: ~~*reson*~~ *reyne*.

345 *con*. So MS, C, A. R, D&A: *can*.

349 *uppight*. So C, A. MS: fully legible only under UV light. R: *upryght*. D&A: *upright*.

350 *hight*. So MS, C, A. R: *n*(?) *hyght*.

355 *is*. C, R: rebound too tightly to read.

358 *seyn*$_2$. So MS, C, A. R: *holden*.

360 *nath*. So MS, C, A. R: *hath not*.

361 *werreourys*. So MS, C. A: *werreores* (*re*$_2$ inserted above the line). R: *werryouris*(?).

365 *shape*. So MS. C: *skape*. A, R: *scape*.

367 *thi foo*. So MS. C: *thy foo*. A: *thy fo*. R: *the so*.

370 *serve*. So C, A, R. MS: *srve*. D&A: *serue*.

374 *Hew*. So MS, C. A: *Hewe*. R: *Smyte*.

376 *is*. So MS, C. A, R: *it*.

to. So MS, C, A. R: *the*.

377 *smyter*. So C, A, R, D&A. MS: fully legible only under UV light.

378 *bonys bite*. So MS, C, A. R: *~~bibe~~ bite*.

380 *thi*. So MS, C. A: *thy*. R: *his*.

383 *kerf*. So MS, C, A. R: *~~arme~~ kerfe*.

384 *in*. So MS, C, A. R omits.

389 *tyme and place*. So MS. C, A: *place and tyme*. R: *place and tym*.

392 *and*. So MS, C, A. R: *in and*.

397 *this al*. So MS, C. A: *this all*. R: *all this*.

398 *doctour*. So MS, C, A. R: *dout(?)*.

400 *preved*. So MS, R. C, A: *previd*. D&A: *proved*.

401 *ye*. So MS, C, A. R: *the*.

402 *had*$_1$. So MS, C, A. R: *~~wold~~ hade*.

406 *amys*. So MS, C. A: *armis* (eye skip to *armys* in line 409, below). R: *armys*.

408 *in*$_1$. So MS, C, A. R omits.

410 *deth*. So MS, C, A. R omits.

418 *ayeyn*. So MS, C, A. R: *agayn*. D&A: *ageyn*.

422 *teche*. So MS, C, A. R: *~~ty~~ teche*.

it. So MS, C, A. R: *for*.

423 *at*$_1$. So MS, C, A. R omits.

425 *the doctour heer*. So MS, C, A. R: *her the doctour*.

436 *craftily*. So MS, A. C: inserted above the line. R: *craftyly*.

440 *on*. So MS, R. C: *oone* inserted in the line. A: *one* (*e* erased).

441 *eek*. So MS, A. C: *~~ek~~ eek*. R: *eke*.

442 *bak*. So MS, C. A: *backe*. R: *~~bla~~ back*.

444 *Cotidian*. So MS, C. A: *Cotidyan*. R: *~~Codi~~ Cotidian*.

451 *Scipio*. So MS, A, R. C: *~~S~~ Scipio*.

455 *may*. So MS, C, A. R: *many*.

458 *are*. So MS. C, A: *ar*. R: *as*.

461 *defende.* So C, A, D&A. MS: *defend* partially legible only under UV light. R: *difende.*

464 *be spende.* So MS, A, R. C: ~~*wyth stonys*~~ *be spende.*

465 *Or ellys thus: save other shot with stonys.* Omitted in R (error).

468 *is good.* So C, A, R. MS: fully legible only under UV light. D&A: *[is] good.*

473 *co-emperour.* So MS, C, A. R: *emperoure.*

475 *lepe.* So MS, C, A. R: ~~*lepe*~~ *lype.*

477 *is.* So MS, C, A. R omits.

481 *LX.* So MS, C. R: *XL.*

is good. So C. MS: fully legible only under UV light. R: *is goode.* D&A: *it hade.*

481–977 *And LX . . . . and refourme.* Omitted in A (lost leaves).

483 *and*$_1$. So MS, C. R omits.

and$_3$. So MS. C, R: *or.*

489 *adayes.* So MS, C. R: *of dais.*

491 *of gere.* MS: *of* ~~*the*~~ *gere.* C, R, D&A: *of the gere.*

495 *an oste it.* So MS. C: *an ost it.* R: *and oft(?) it.*

496 *feet.* So C, R. MS: *f . . .* partially legible only under UV light. D&A: *fvlly.*

498 *safly.* So MS, C. R: *saufly.* D&A: *saftly.*

499 *never.* So MS, R. C: ~~*never*~~ *never.*

500 *ha.* So MS, C. R: *a.*

503 *a.* So MS, C. R omits.

504 *deep.* So MS, C. R: ~~*b*~~ *depe.*

505 *goode.* So MS. C: ~~*gode*~~ *gode.* R: *good.*

508 *May.* So MS, C. R: ~~*May*~~ *May.*

515 *uch.* So MS, C. R: *iog(?).*

517 *to make.* So C, R, D&A. MS: *to* ~~*be*~~ *make.*

522 *Tho.* So MS, C. R, D&A: *The.*

523 *picche.* So MS, C. R: *put he.*

524 *and dragonys.* So C, D&A. MS: fully legible only under UV light. R: *and dragons.*

527 *and.* So MS. C, R: *in.*

531 *ther oppresse.* So MS. C, R: *oppresse ther.*

535 *IX.* So MS, R. C: inserted in the line.

537 *werk.* So MS, C. R: *dike.*

539 *is.* So MS, R. C: ~~*is*~~ *is.*

542 *as is best on either side.* So MS. C: ~~*and*~~ *as is best on every side.* R: *on every side as best is.*

550 *redy.* So MS, C. R omits.

557 *forth hem calle.* So MS, C. R: *call hem forth.*

563 *In.* So MS, C. R: *Ther.*

565 *in fight.* So MS. C: *in sight.* R: *sure.*

566 *frequentaunce.* So MS, C. R: *frequent a vice.*

571 *atwynne.* So MS, C. R: ~~*thyme*~~ *twynne.*

577 *us.* So MS, C. R: *we.*

598 *saladed.* So MS, C. R: *saltid.*

599 *her.* So MS, C. R omits.

607–13 *Ereithre ege . . . . or harmys.* Omitted in R (likely error).

614 *Th'electioun and exercise.* So MS, C. R: *Excercise and theleccion.*

616 *uch.* So MS. C: *uche.* R: *everyche.*

620 *not.* So MS, C. R: ~~*to*~~ *not.*

Before 621 *Electrix ita . . . . secunda subit.* Latin omitted in C, R.

621 *part.* So MS, C. R, D&A: *parte.*

626 *withoute.* So MS, C. R: *with non* (*ith non* inserted above the line).

634 *wit.* So MS. C: ~~*wil*~~ *wyt.* R: *witt.*

643 *armys stont.* So MS, C. R: *manys armys.*

644 *on*$_2$. So MS, C. R: *y*(?)*on.*

646 *eke.* So MS, R. C: ~~*ek*~~ *eke.*

648 *havyng.* So R, D&A. MS: *having havyng* (uncancelled dittography). C: *having.*

664 *Lith.* So MS, C. R: *Hath* (eye skip to *hath* in line 666, below).

th'. So MS. C: inserted above the line. R omits.

669 *applie.* So MS, C. R: *to plie.*

674 *The.* So MS, C. R: *This.*

678 *Al.* So MS. C: *At.* R: ~~*But*~~ *At.*

682 *Wyis.* So MS, C. R: *Whys is.*

690 *by.* So MS, C. R omits.

691 *honoure is*. So MS. C: rebound too tightly to read. R: *honour*. D&A: *honour is*.

694 *uch*. So MS, C. R: *with*.

ympendent. So MS. C: ~~*ymped*~~ *ympedent*. R: *impenduet*(?).

697 *hym*. So MS, C, R. D&A: *him*.

698 *Criste Jhesus*. So MS. C, R: *Jhesus Criste*.

707 *Cohors the*. So C. MS: *Cohors of*(?) *the*. R: *To hors*.

708 *us, good*. So MS. C: *us,* ~~*god*~~ *good*. R: *is, good*.

709 *number*. So MS, C. R: *the nombre*.

710 *cohors*. So MS. C, R: *cohort*.

720 *the*$_1$. So MS, C. R: *is the*.

722 *it*. So MS. C: *hit*. R omits.

723 *V*$_1$. So MS, C. R omits.

726 *in*. So MS, R. C omits.

727 *espoyre*. So MS. C, R: *espoire*. D&A: *espeyre*.

729 *discrive*. So MS. C: *descrive*. R: ~~*st*~~ *distryve*.

731 *as*. So MS, C. R: *also*.

738 *yet*. So MS, C. R omits.

749 *XXX therto*. So MS, C. R: *therto XXX*.

751 *roote*. So MS, C. R: *r*(?) *rote*.

756 *retourne*. So MS, C. R: *is retorne*.

760 *us*. So MS, C. R: *be*.

762 *And*$_1$. So MS, C. R omits.

send. So MS, C. R: ~~*send*~~ *sent*.

768 *An*. So MS, C. R: ~~*E*~~ *An*.

770 *ther were of those*. So MS. C: *there were of those*. R: *of tho ther werre*.

773 *he*$_1$. So MS, C. R omits.

775 *exercise*. So MS, C. R: *and excercise*.

786 *resoun*. So C, R. MS: *reasoun*.

787 *thei*. So MS. C: *they*. R omits.

791 *and*. So MS, C. R: *he*.

ye$_1$. So MS, C. R: *the*.

	ye$_2$. So MS, C. R: *thei.*
805	*make.* So MS, C. R: *ayake.*
807	*th'egil forto arise.* So MS, C. R: ~~*e*~~ *the egle is forto arise.*
825	*an.* So MS, C. R: *an~~d~~.*
826	*theim; and.* So MS. C: *theym; and.* R: *the.*
	it. So MS, R. C omits.
830	*ofte.* So MS, C (or *oste*?). R: *hoste.*
833	*hem.* So MS, C. R omits.
837	*that.* So MS, C. R: *to that.*
841	*forto.* So MS, C. R: *therto.*
845	*For.* So MS, C. R omits.
	th'ordinayrys. So MS. C: *thordynarris.* R: ~~*or*~~ *thordynaris.*
852	*consulys.* So MS. C: *consulis.* R: ~~*consuls*~~ *conssellis.*
854	*th'estatys.* MS: *thestatys.* C, R: *thestatis.* D&A: *the statys.*
857	*poerys.* So MS. C: *perys.* R: *peris.* D&A: *Peerys.*
865	*and*$_2$. So MS, C. R: *of.*
866–977	*Of him . . . . and refourme.* Omitted in R.
869	*or.* So MS. C: *of.*
884	*the.* So MS. C omits.
901	*Forgys.* MS: *Forergys.* C: *Forgis.* D&A: *Foregys.*
902	*make.* So C, D&A. MS: *male.*
927	*inscribed.* So MS, C. D&A: *inscrived.*
943	*fetured.* So MS. C: *statured.*
951	*feat.* So MS. C: ~~*fight*~~ with *fayt* inserted above the line.
974	*al.* So MS. C: inserted above the line.
978–1026	*Tercia bellatrix . . . . and bataile.* Omitted in C, A, R (anti-Lancastrian view).
992	*Goldon.* So MS. D&A: *golden.*
1035	*his.* So MS, C, A. R: *a.*
1037	*cohortatioun.* So MS, C. A: ~~*ch*~~ *cohortatioun.* R: *cohortacion.*
1046	*groweth.* So MS, C, A. R: ~~*goo*~~ *with.*
1056	*hath.* So MS, C, A. R: ~~*had*~~ *I hath.*

1063 *slough*. So C, A, R, D&A. MS: *shough*.

1066 *of vitaile*. So MS, A. C: ~~*among*~~ *of vitaile* (eye skip to *amonge* in line 1065, above). R: *of bataile*.

1067 *turne*. So MS, C, A. R: ~~*ton*~~ *turne*.

1073 *chivalerys*. So MS. C: *chyvalry*. A: *chivalry*. R: *chevalery*.

1075 *Is*$_1$. So MS, C, A. R: *As*.

1079 *no*. So MS, C, A. R: *of no*.

1102 *ever it*. Ful. So MS. C: *ever hit*. Ful. A: *ever hit*. Full. R: *it ever. Full Full*.

be. So MS, C, R. A: ~~*by*~~ *be*.

1114 *The colde fyer of indigence is hoote*. Omitted in R (error).

is. So MS, C. A: *his* (*h* erased).

1115 *theron*. So MS, C, A. R: *ther~~to~~on*.

1119 *the*. So MS. C, A: *thy*. R: *your*.

1123 *tak*. So MS. C, A: *take*. R: ~~*e*~~ *take*.

1127 *progressioun*. So C, A, R, D&A. MS: *progeressioun*.

1130 *within, and enmytee abowte*. So MS. C, A: *withinne, and enmytee abowte*. R: *withinne withinne* ~~*then was withoute*~~ with *and emite abowte* inserted above the line (eye skip to *then was withoute* in line 1131, below).

1131 *is then*. So MS, C, A. R: *then was*.

1132 *withoute an*. So MS, C, A. R: *with*.

1133 *sle*. So MS, C, A. R: *fle*.

1134 *Forthi*. So MS. C, A: *forthy*. R: *For this*.

1140 *to strengthes*. So MS. C: *to strenghis*. A: ~~*of*~~ *strenghis* with *to* inserted above the line. R: *strenghtis* (?).

1141 *lyves*. So MS, C, A. R: ~~*lv*~~ *lives*.

1161 *swere grete, ye*. So MS. C: *grete swere ye*. A: *grete swer*(?) *ye*. R: *grete swere the*.

1162 *ye*. So MS, C, A. R: *the*.

1166 *preef*. So MS, C, A. R: *preff*. D&A: *pref*.

1168 *opon*. So MS, C, A. R omits.

1170 *Com*. So MS, C, A, R. D&A: *cum*.

1174 *Sumtyme*. D&A mistakenly mark this as the folio start; it begins at line 1167.

1182 *wer*. So MS, C, A. R: *had*.

1185 *hoom*. So MS. C: *hom*. A: *home*. R: *homward*.

sers. So C, A, R. MS: *srs*. D&A: *sirs*.

1188 *when*. So MS. C, A(?), R: *wher*.

1189 *his*. So MS, C, A. R omits.

1194 *tech*. So MS. C, R omit. A: *teche* inserted above the line.

1199 *a*. So MS, C, A. R: *many a*.

men. So MS, C, A. R: ~~*to and me*~~ *men*.

1200 *a myghti*. So MS. C, A: *myghty*. R: *myghtyly*.

1206 *a*. So MS, C, A. R: *the*.

1209 *he*. So MS, C, A. R: ~~*the*~~ *he*.

1210 *is*. So MS. C, A, R: *of*.

1211 *is*. So MS, C, A. R omits.

1217 *the*. So MS, C, A. R omits.

1228 *hym*. So MS, C, A. R omits.

1235 *rejecte*. So MS, C. A: *reject*. R: *j*(?) *rejecte*.

1236 *to*. So MS, C, A. R omits.

1241 *folk*. So MS, C. A: *folke*. R omits.

1244 *the*. So MS, C, A. R: *this*.

1246 *deme*. So MS, C, A. R: ~~*dem*~~ *dem*.

1248 *Yet*. So MS, C. A: *Yit*. R: *It*.

1249 *oost*. So MS, C, A. R: *e*(?) *oste*.

1255 *sikerly*. So MS, A. C: *sykerly*. R: *surely*.

1260 *evermore*. So MS, C, A. R: *overmore*.

1261 *of*. So MS, C, A. R: ~~*in*~~ *of*.

1267 *pipe*. So MS, C, A. R: *n*(?) *pipe*.

1269 *the*. So MS, C, R. A: ~~*tho*~~ *the*.

1275 *place*. So MS, C, A. R: *space*.

1282 *blewe up*. So MS. C: *belleth* with *vel blew up* inserted above the line. A: *bellith* with *vel blow up* inserted above the line. R: *belleth*.

1285 *the*. So MS, C, R. A: inserted above the line.

1292 *his*. So MS, C, A. R: ~~*de*~~ *his*.

1303 *mad*. So MS, C. A: *made*. R: ~~*make*~~ *made*.

1309 *is*. So MS, A. C: ~~*is*~~ *is*. R: *ther is* (eye skip to *ther is* in line 1308, above).

resistence. So MS, C, A. R: *the resistiance*.

1313 *theron oonly ther bowe*. So MS. C, A: *oonly theron ther bowe*. R: *oonly theron bowe theron* (*on* inserted above the line).

1316 *by*. So MS, C, A. R omits.

1321 *journal*. So MS, A. C: ~~*joo jo*~~ *journal*. R: *jornall*.

1323 *Wheryn have he*. So MS. C: *Theryn have he*. A: *Therin have he*. R: *Therin hath been*.

1328 *is*. So MS, C, A. R: ~~*as to*~~ *is*.

wise. So MS, C, A. R: ~~*a*~~(?) *wise*.

1333 *on*. So MS, C, A. R: ~~*w*~~ *on*.

1335 *of*. So MS, C, A. R omits.

1339 *Whethour*. So MS, C. A, R, D&A: *Whethor*.

1341 *and*$_2$. So MS, A, R. C: *of*.

1342 *to*. So MS, C, A. R: *a*.

1344 *perfit*. So MS, R. C: *perfyt*. A: *parfyt*. D&A: *parfit*.

1345 *he*. So MS, C, A. R: *to*.

1346 *forthi*. So MS. C, A: *forthy*. R: *therfor*.

1348 *simpilnesse*. So MS. C: *symplenesse*. A: *simplenesse*. R: *symples*.

1349 *goth*. So MS, A, R. C: ~~*th*~~ *goth*.

1350 *be*. So MS, A, R. C: ~~*be*~~ *be*.

1354 *the*. So MS. C, R: *thy*. A: *thi*(?).

1358 *swere*. So MS, C. A: *sweru*(?). R: *sure*.

1361 *wightest*. MS, D&A: *wightiest*. C: *wyghtest*. A: *wyghtist*. R: *wightist*.

1363–2300 *In a maner . . . . and mon*. Omitted in A (lost leaves).

1377 *withal*. So MS, C. R: *with that*.

1378 *the*. So MS. C, R: *thy*.

falle on. So MS. C: *fal on*. R omits.

1379 *it*. So MS, C. R: ~~*to*~~ *it*.

1380 *abide*. So MS, C. R: *to abide*.

1382 *beforn*. So MS, C. R: *aforne*.

1389 *make*. So MS, R. C: ~~*may*~~ *make*.

1401 *it*. So MS. C: *it* corrected from *is*. R: *that*.

1402 *And to bewar that slough viage or flood is*. This line mistakenly written at the end of the stanza in MS, corrected with marginal notation.

to. So MS, C. R omits.

1411 *o*. So MS, C. R omits.

1413 *An*. So MS, C. R, D&A: *And*.

1418 *her*. So MS. C: ~~*w*~~ *her*. R: *here*.

1423 *of*. So MS, C. R: *therof*.

1425 *over-hede*. So MS. C: *over-~~hod~~-heed*. R: *over-hed*.

1428 *is*. So MS, R. C: *is is* (uncancelled dittography).

1430 *Ye*. So MS, C. R: *the*.

1431 *wher*. So C, R. MS: *where*.

by$_2$. So MS, C. R: inserted above the line.

1433 *It*. So MS, C. R omits.

1441 *festeying calle*. So MS, C. R: *festyeing call*. D&A: *festeyng call*.

1452 *men*. So MS, C. R omits.

1456 *therof to breke*. So MS. C: *therof to breek*. R: *to brek therof*.

plye. So MS. C: *plie*. R: ~~*pls*~~(?) *plie*.

1457 *Benethe*. So MS. C: *Beneth*. R: *Be~~nn~~neyth*.

1462 *it*$_1$. So MS, C. R omits.

1467 *ynough it is*. So MS, C. R: *it is inough*.

1470 *better*. So MS, C. R omits.

1472 *With*. So MS, R. C: ~~*An*~~ *With* (eye skip to *And* in line 1473, below).

1473 *tabil*. So MS, C. R: *cable*.

1479 *up*. So MS, C. R: ~~*u*~~ *up*.

1480 *there*. So MS. C: *ther*. R: *there~~in~~*.

1481 *vitaile*. So MS. C: *vitayle*. R: *bataile*.

1482 *Now*. So MS, C. R: ~~*Now*~~ *Now*.

1490 *ayer, and*. So MS. C: *ayre, and*. R: *or of*.

1491 *Holsom be*. So MS. C, R: *Be holsome*.

1493 *For man and hors, and woode ynough at honde.* This line mistakenly written at the end of the stanza in MS, corrected with marginal notation.

1503 *and.* So MS, C. R: ~~*and herde*~~ *and.*

1504 *Up.* So MS, C. R: ~~*Up*~~ *Up.*

1506 *fote.* So MS. C: *foot.* D&A: *foote.*

1507 *be, not herfore.* So MS. C: *be, not therfore* (*t* struck out). R: *it be therfore.*

1508 *Turf.* So MS, C. R: ~~*Therfor*~~ *Turf.*

1512 *to dwelle heryn.* MS: *dwelle* written over an erasure. C: *to dwell heryn.* R: *herin to* ~~*stand*~~ *dwell.*

1518 *dich.* So MS, C. R: *dike.*

1527 *it.* So MS, C. R omits.

1530 *werk.* So MS, R. C: *castell werk* (eye skip to *castel* in line 1528, above).

1534 *not.* So MS, C. R omits.

perfit. So MS. C: *parfyt.* R, D&A: *parfit.*

1536 *himself.* So MS. C: *hymself.* R: *hyme~~self~~selff* (*l* inserted above the line).

1539 *on foote.* So MS, C. R omits.

1542 *for.* So R. MS, C: inserted above the line.

1544 *thei.* So MS. C omits. R: *it.*

nothing. So MS, C. R: *onthing.*

1546 *set.* So MS, C. R: *fexe.*

1551 *is had to picch.* So MS. C: *is to picch.* R: ~~*is picch*~~ *is had to picch* (*is had* inserted above the line).

1554 *deperted.* So MS, R. C, D&A: *departed.*

1558 *horsed men.* So MS, C. R: *horsmen.*

1568 *for.* So MS, C. R omits.

1572 *Ye.* So MS, C. R: *The.*

1579 *on.* So MS, C. R: *uppon.*

uppon. So MS, C. R: *agayn.*

1582 *prince.* So MS, C. R: ~~*prnce*~~ *prince.*

1585 *him.* So MS. C: *hym.* R omits.

1597 *her*$_2$. So R, D&A. MS, C: *he.*

1600 *or.* So MS, C. R: *than.*

1601 *the.* So MS, C. R: inserted above the line.

1603 *Wher*. So MS, C. R: *Whethir.*

1610 *the*. So MS, C. R: inserted above the line.

1611 *balaunce*. So MS, C. R: *a ballaunce* (*a* inserted above the line).

1612 *is*$_2$. So MS, C. R: *to.*

1619 *that*. So MS. C, R: *and.*

1620 *ner*. So MS, C. R, D&A: *nor.*

1624 *Receyved*. So MS, C. R: *Reytayned.*

1625 *croune*. So C, R. MS: *c~~o~~roune.*

1634 *alle men*. So MS. C: *al men*. R: ~~*alle*~~ *all men* (*all* inserted above the line).

1651 *of*. So MS, C. R omits.

1662 *ar*. So MS. C, R, D&A: *or.*

1684 *of*$_2$. So MS, C. R omits.

1685 *til now was ther no werre*. So MS, C. R: *ther was no warre till now.*

1689 *their*. So MS, C. R: *yow.*

1693 *the*. So MS, R. C omits.

1699 *mad newe*. So MS. C: *newe~~e~~ mad*. R: *made newe.*

1708 *o Lord, this day*. So MS. C: *o Lorde, this day*. R: *this day, o Lorde.*

1715 *prince is*. So MS. C: *prynce is*. R: *princis.*

1719 *let*. So MS. C: *lete*. R: ~~*so*~~ *let.*

1721 *citee*. So MS, C. R: *in cite.*

1724 *to*. So MS, C. R: ~~*to*~~ with *to* inserted above the line.

1732 *hem*. So MS, C. R omits.

1739 *bataile*. So MS. C: ~~*b*~~ *bataile*. R: *batayle.*

1740 *yet*. So MS, C. R omits.

1744 *thei ha be shake*. So MS. C: *they have shake*. R: *that thay have shall.*

1751 *siker*. So MS. C: *syker*. R: *sure.*

1752 *comyng*. So MS, C. R: *com comyng* (uncancelled error).

1753 *a*. So MS, R. C omits.

1757 *to*. So MS, C. R omits.

1761 *cleef*. So MS. C: *chef*(?). R: *chefe.*

1763 *Ereither*. So MS, C. R: ~~*E*~~ *Arayether.*

1764 *Ereither*. So MS, C. R: *Other*.

1765 *wynde to fight*. So MS. C: *to fight wynd*. R: *to fight wynd*.

1767 *so*. So MS, C. R: ~~*so*~~ *so*.

1768 *they her*. So MS. C: *he his*. R: *he is*.

1770 *And*$_1$. So MS, R. C: ~~*Wil*~~ *And* (eye skip to *wil* at line 1773, below).

1773 *as*. So MS, R. C: *a as* (uncancelled error).

1775 *he*. So MS, C. R: *the*.

1783 *is*. So MS. C, R: *it*.

1784 *foon*. So MS, C. R: *soon*.

1790 *a principaunt*. So MS, C. R: ~~*a per*~~ *a principaunt* (*a* inserted above the line).

1802 *a*. So C, R, D&A. MS omits.

pace. So MS, C. R: ~~*foot*~~ *pace*.

1803 *DC*. So C, R(?). MS, D&A: *CC*. See Explanatory Note to this line.

1807 *so*. So MS, R. C: ~~*s*~~ *so*.

1816 *hem*. So MS, C. R: *ther*.

1817 *reere*. So MS. C: *rere*. R: *ver*.

1825 *grete*. So MS. R: *gret*. C: ~~*g*~~ *grete*.

1826 *is sette*. So MS, C. R: *sett is*.

1832 *Wight*. So MS, C. R: *With*.

1835 *keste*. So MS, R. C: ~~*caste*~~ *keste*.

1851 *colveryne*. So MS, C. R: ~~*on*~~ *colveryne*.

1853 *faughte*. So MS. C: *faughte* (*e* inserted above the line). R: *faught*.

1864 *reperatioun*. So MS. C, D&A: *reparatioun*. R: *reparacioun*.

1868 *to fight — it to se*. So MS (with punctuation). C: *the sight it is to se*. R: *the sight it is to see*.

1873 *veer*. So MS. C, R: *beer*.

1883 *double londe*. So MS, R. C: ~~*threfold*~~ *space* with *double* inserted above the line.

1886 *podisme*. So C, R. MS, D&A: *pidisme*.

1891 *and*$_2$. So MS, C. R: *or*.

1892 *rangis*. So MS. C: *eggis* with *vel raungis* inserted above the line. R: *eggis*.

1894 *is brede*. So MS. C: ~~*brede*~~ *is brede*. R: *is bred*.

1905 *be*. So MS, R. C: *b be* (uncancelled error).

1924 *on.* So MS, R. C: ~~*of*~~ *on.*

1927 *sharre.* So MS. C, R: *sharpe.*

1931 *or wynge.* So MS (inserted above the line in different ink), C, R.

1932 *hem.* So MS, C. R: ~~*of*~~ *hem.*

1939 *mad.* So MS, C. R omits.

1948 *ye.* So MS, C. R: *the.*

1950 *fier.* So MS. C: ~~*foyr*~~ *fier.* R: *fire.*

1961 *that.* So MS, R. C: ~~*a foote*~~ *that* (eye skip to *o foote* in line 1962, below).

1962 *t'abide.* So MS. C: *taby.* R: *to abide.*

1968 *God.* So MS, C. R: ~~*good*~~ *Gode.*

1980 *it.* So MS, R. C: *is.*

1984 *It.* So MS, C. R: *Yet.*

1990 *his.* So MS, R. C: *his* (*h* inserted above the line).

1991 *Their.* So MS, C. R: *Ther.* D&A: *Then.*

yfere. So MS, C. R: ~~*jo*~~ *fere.*

so. So MS, C. R omits.

horribil. So MS, C. R: ~~*horibbe*~~(?) *horible.*

1992 *incredibil.* So MS, C. R: *credible.*

2004 *hem.* So MS, C. R omits.

2005 *on eny.* So MS. C: *in eny* (*meny*?). R: *many.*

2011 *Omnipotens.* So MS, R. C: *Omnipotens* ~~*Lord*~~.

2012 *loveth.* So MS, R. C: ~~*loved*~~ *loveth.*

sapience. So MS, C. R: ~~*dispence*~~ *sapence* (eye skip to *dispence* in line 2013, below).

2013 *justice to.* So MS, C. R: *is justice and.*

2023 *noo.* So MS. C, R, D&A: *no.*

2025 *fle.* So MS, C. R: *flee.*

2026 *if.* So MS. C: *yf.* R omits.

2041 *and.* So MS, C. R: ~~*is*~~ *and.*

2043 *also fewe.* So MS, C. R: *with few also.*

2046 *Now moste.* So MS. C: *Now most.* R: *Most now.*

2057 *other.* So MS. C: *their.* R: *either.*

2059 *myghtiest*. So MS, C. R: *the myghtiest*.

2063 *are*. So MS. C: *ar*. R: *as*.

a. So MS. C, R: *and*.

2071 *hold it*. So MS. C: *holde ~~hyth~~ hit*. R: *it*.

2075 *circuyte*. So MS, C. R: *the circute*.

bringe. So MS. C: *bryng*. R: *~~a~~ bryng*.

2077 *do*. So MS, C. R omits.

2080 *to*. So MS, C. R omits.

bede. So MS, R. C: *~~boode~~ bede*.

2092 *al putte*. So MS, C. R: *put all*.

2093 *D*. So MS, C. R omits.

2099 *and*$_2$. So MS, C. R omits.

2108 *his*. So MS, C. R: *thy*.

2112 *retrete*. So C, R, D&A. MS: *retretrete*.

2139 *turneth*. So MS. C: *turnyth*. R: *knyght*.

2152 *the*. So MS, C. R: *~~te~~ the*.

2160 *resoun*. So C, R. MS: *reasoun*.

2165 *everywhere*. So MS. C: *~~ever~~ everywhere*. R: *everywher*.

2170 *do*$_3$. So MS. C: *doos*. R: *doo*.

2175 *to fight a knyght unexercised*. So MS. C: *to fight a knyght ~~unexercise~~ unexercised*. R: *a knyght to fight unexcercissid*.

2178 *untrested*. So MS, C. R: *understand untrested*.

2182 *to fight then lengthe*. So MS. C: *to fight then lengh*. R: *thyn fight in lengh*(?).

2183 *in stoor to have*. So MS, C. R: *to have in storre*.

2185 *Fende*. So MS, C. R: *Fend*. D&A: *Sende*.

2186 *wightly*. So MS, C. R omits.

2197 *Aforn the*. So MS, C. R: *~~Afro~~forne the*. D&A: *Aforn thi*.

2202 *oon*. So MS, C. R omits.

2204 *cleef*. So MS, C. R: *chefe*.

Espie. So MS, C. R: *To espye*.

2207 *for*. So MS, C. R: *forth*.

2209 *stele*. So MS, C. R: *~~away~~ stell*.

2215 *not to*. So C, MS (*to* inserted above the line). R: *nor to*.

2222 *I here*. So MS. C: *here I*. R: *her I*.

Before 2224 *Ultima pars . . . . ornat opus*. Latin omitted in C, A, R.

2226 *morthereer*. So MS. C: *morthe~~t~~reer*. R: *murtherer*.

2235 *noone*. So MS, C. R: *~~v~~ noone*.

2241 *it*. So MS, R. C: inserted above the line.

2244 *assege*. So MS. C: *~~assege~~ assege*. R: *asege*.

see. So MS. C: *se*. R: *so*.

2249 *shal be*. So MS, C. R omits.

2250 *cragge*. So MS, C. R: *oragge*.

2255 *lengthe*. So C. MS: *lenghe*, followed by D&A. R: *lynghe*.

2262 *or*. So C, R, D&A. MS: *or or* (uncancelled dittography).

2265 *kest*. So MS, C. R: *~~kest~~* with *kest* inserted above the line.

2267 *an*. So MS. C: *and in*. R omits.

2274 *as*. So MS, C. R omits.

2275 *Me*. So MS, C. R: *We*.

2277 *ram*. So MS, C. R: *rayne*.

2278 *cruste*. So MS, C. R: *couste*.

2286 *sle*. So MS. C: *sle* corrected from *fle*. R: *fle*.

2291 *rathest*. So MS, C. R: *right*.

2297 *cratys*. So MS. C: *~~cratis~~* with *rackys* inserted above the line. R: *rackys*.

2299 *crate*. So MS. C: *~~crate~~* with *racke* inserted above the line. R: *racke*.

2306 *within*. So MS. C, A: *in with*. R: *withinne*.

2312 *that*. So MS, C, A. R omits.

2313 *vitaile*. So MS, C. A: *vittaile*. R: *~~ve~~*(?) *victaile*.

2317 *that*. So MS, C, A. R omits.

2322 *the*. So MS, C, A. R omits.

saddest. So MS, R. C: *sadde*. A: *saddest* (*st* inserted in line).

2325 *hegh*. So MS. C: *hegth*. A: *heght*. R: *hiegst*.

Before 2329 MS: *And yf theire be any yon* written in top margin below running title.

2329 *is.* So MS, C, A. R omits.

2331 *engyne*$_2$. So MS, C, A. R omits.

2333 *over.* So MS, C, A. R: ~~*of*~~ *on.*

2353 *make.* So MS, C, A. R: *make make* (uncancelled dittography).

2359 *litil.* So MS, C, A. R omits.

2360 *fette.* So MS, C. A: *sette.* R: *sett.*

2363 *doubte.* So MS. C, A: *doute.* R omits.

2366 *citee nys.* So MS, C, A. R: ~~*cite nes*~~ *cite nys.*

2370 *forth.* So MS, C. A: *furth.* R omits.

2375 *eke.* So MS, A. C: *ek.* R omits.

2376 *Lo alle these wil this citee assaile.* MS: *alle* written over an erasure. C: *Lo* ~~*al these wil*~~ *this citee wil alle these assaile* (*wil alle these* inserted above the line). A: *Lo this citee wil all these assaile.* R: *To this cite will all thes assaille.*

2379 *shelled snail.* So MS. C, A: *snayle shelled.* R: *snaile shelled.*

2380 *it.* So MS, C, A. R omits.

2381 *hornys.* So MS, C, A. R: *the hornys.*

2384 *their.* So MS, C, A. R omits.

2388 *theron fier.* So MS. C: *fyer theron.* A, R: *fier theron.*

2391 *iron-steeled stronge.* So MS, C. A: *yron-steeled* ~~*stor*~~ *stronge.* R: *irone-steled stronge.*

2396 *have.* So MS, C, A. R: *be.*

it hooke. So MS. C: *it hooke* (*o*$_2$ inserted above the line). A, R: *if it hooke.*

2399 *and*$_2$. So MS, C, A. R omits.

2401 *rush.* So MS, A. C: ~~*rush*~~ *rush.* R: *rushe.*

2403 *oon.* MS, C, A, R, D&A: *oo.*

2404 *a trusse.* So MS, C, A. R: *attonsse.*

2406 *kecch it.* So MS. C: *kecchith.* A: *kecchith* with *it* inserted above the line. R: *kecchith it.*

2411 *as.* So MS, C, A. R omits.

2417 *of*$_1$. So MS, C, A. R omits.

2422 *every.* So MS, R. C: *hevy.* A: *every* (*h* scraped from beginning and *er* inserted above the line).

2426 *this.* So MS, C, A. R omits.

2432 *covere.* So MS. C, A: *overe.* R: *over.*

2434 *ynowe*. So MS, C, A. R: ~~*an*~~ *inowe*.

2440 *undir*. So MS, C. A: *under*. R: ~~*other*~~ *undre*.

2445 *III*. So MS, C, A. R: ~~*to go*~~ *III*.

thei thise demene. So D&A. MS: . . . . *e* partially legible only under UV light. C: ~~*they*~~ *they this demene*. A: *they this demene*. R: *thay this demene*.

2455 *eke is good*. So MS. C, A: *ek is good*. R: *is good eke*.

2456 *eny*. So D&A. MS: written over an erasure. C: *al*. A, R: *all*.

2459 *hem*. So MS, C, A. R omits.

2469–2580 *Present perile . . . . foo bewere*. Omitted in A (lost leaves).

2470 *in a doubte*. So MS. C: *in a doute*. R: *to devote*.

2471 *An*. So MS, C, R. D&A: *And*.

2479 *in*. So MS, C. R: ~~*v*~~ *in*.

2488 *also*. So MS, C. R omits.

2489 *it*. So MS. C, R omit.

2491 *A fallary, a shafte is of the same*. This line is mistakenly written at the end of the stanza in R. See the Introduction, pp. 12–13.

2492 *Thorgh*. So MS, C. R: ~~*A fallery*~~ *Throwe* (*Throwe* inserted above the line).

2494 *this*. So MS, C. R: *the*.

2506 *ther*. So MS, C. R omits.

2511 *ye*. So MS, C. R: *the*.

2514 *throgh*. So MS. C: *thorgh* (*r* inserted above the line). R: *thorrogh*.

2517 *sleth*. So MS, C. R: *fleeth*.

2518 *concavitee*. So MS, C. R: ~~*conv*~~ *concavite*.

2519 *whelis*. So MS (added in margin at end of the line in darker ink), R. C: *wheelys*.

2520 *goth*. So MS, C. R: ~~*goth*~~ *goth*.

2522 *dolour*. So MS, C. R: ~~*dolur*~~ *dolour*.

2528 *thei*. So MS. C: *they*. R omits.

2535 *adoun*. So MS, C. R omits.

2538 *other*. So MS, C. R: *ther*.

2551 *or*. So MS, C. R: *and*.

onys. So MS. C: *ony*. R: *any*.

2557 *Or*. So MS, C. R: ~~*Oa*~~(?) *Or*.

2565 *out*. So MS. C: *oute*. R omits.

2570 *thenne*. So MS. C: *then*. R omits.

2571 *we*. So MS, C. R: *we we* (uncancelled dittography).

2572 *Aslepe*. So MS. C: *Asleep*. R: ~~*Asl*~~ *Aslepe*.

2576 *everychon*. MS: *everychon~~e~~*. C, R: *everychone*.

2583 *postis*. So MS, A, R. C: ~~*b*~~ *postis*.

2585 *wel*. So MS, C, A, D&A. R: *wele*.

2596 *wise*. So MS, C, A. R: *good*.

2598 *wake*. So MS, C. A, R: *make*.

2608 *on other*. So MS, C, A. R: *other*. D&A: *an other*.

2617 *Good*$_{1}$. So C, A, R, D&A. MS: fully legible only under UV light.

2618 *But*. So C, A, D&A. MS: fully legible only under UV light.

2619 *stuf*. So MS. C: *stufe*. A, D&A: *stuffe*.

2622 *the*. So C, A, R, D&A. MS: fully legible only under UV light.

2625 *good*. So C, A, R, D&A. MS: fully legible only under UV light.

2626 *nayles . . . bras*. So D&A. MS: fully legible only under UV light. C: *naylis . . . bras*. A: *nailis . . . bras*. R: *nailles . . . bras*.

2628 *kanker*. So D&A. MS: fully legible only under UV light. C, A, R: *canker*.

2634 *pith*. So C, A, R, D&A. MS: *p.th* partially legible only under UV light.

2635 *XV*. So MS, C, A. R: *How*.

2653 *myghtier*. So MS, C, A, R. D&A: *mightier*.

2655 *sturne*. MS: *s . . . ne* partially legible only under UV light. C, A, D&A: *storne*. R: *sterne*.

2657 *the*. So MS, C, A. R omits.

or. So C, A, R, D&A. MS: fully legible only under UV light.

2661 *that not*. So MS, C, A. R: *that*. D&A: *thei not*.

2662 *victory*. So C, A, R, D&A. MS: fully legible only under UV light.

2663 *marynere*. So D&A. MS: *ma.ynere* partially legible only under UV light. C, A: *maryneer*. R: *mariner*.

2664 *unnapere*. So MS, C. A: *unnapeer*. R: *unrape*.

2669 *arowe*. So MS, C, R. A: ~~*abowe*~~ *arowe*.

2671 *thei amonge*. So MS. C, A: *they among*. R: *among they*.

2673 *Sir*. So C, A, R. MS: inserted above the line.

2674 *tan*. So MS, R. C omits. A: inserted in the line.

2676 *of the North*. So C, A, R, D&A. MS: *of North*.

2681 *lifte*. C, A: *lefte*. R: *left*. D&A: *lift*. MS: *right*.

2683 *the . . . when*. So C, A, R, D&A. MS: fully legible only under UV light.

2687 *of*. So MS, C, A. R: *that*.

nature. So MS, C. A: *natur*. D&A: *nater*.

tranquylle. So MS. C, A: *tranquille*. R: ~~*a*~~ *tranquille*.

2688 *lene*. So MS, A, R. C: ~~*Heve*~~ with *lene* inserted above the line.

2689 *al*. So MS. C: inserted above the line. A, R: *all*.

2690 *rore*. So C, A, R. MS: fully legible only under UV light. D&A: *ror*.

2691 *this*. So MS. C: *is*. A: ~~*is*~~ with *this* inserted above the line. R: *is thus*(?).

2692 *me*. So C, A, R. MS: inserted above the line.

hert it. So MS, A. C: ~~*her*~~ *hert it*. R: *hert* ~~*it*~~ *it*.

2699 *Husht*. So MS, C, A. R: ~~*Hushe*~~ *Hush*.

2704 *in*. So C, A, R. MS: inserted above the line.

2709 *affrays*. So MS, C. A: *affrayis*. R: *affrais*. D&A: *affrayes*.

2712 *intractabil*. So MS, C, A. R: *intraitabill*. D&A claim MS reading is *mutabil* (p. 99n2712).

2717 *noo*. So MS. C, A, R, D&A: *no*.

2720 *plesaunt*. So MS, C, A. R omits.

2722 *Ide*. So MS, C, A. R: ~~*Id*~~ *Ide*.

2724 *remembre*. So MS. C: ~~*re*~~ *remembye*. A, R: *remembr*.

2726 *meanewhile*. So MS. C, A: *meanwhile*. R: ~~*whill*~~ *meaynwhill*.

2730 *see*. So MS, C, A. R: *the see*.

2733 *or ride*. So MS, C. A: ~~*and*~~ *or ride* (*or* inserted above the line). R: *nor ride*.

2734 *a*. So MS, C. A: inserted above the line. R omits.

2735 *is*$_2$. So MS, C, A. R omits.

2739 *o londe*. So MS, C, A. R: *to long*.

2740 *leyt*. So MS, C. A: *leit*. R: *slet*.

2741 *goth*. So MS. C omits. A: inserted above the line. R: *gooth*.

2744 *sollennytee*. So MS. C, A: *solempnytee*. R: *solempnyte*. D&A: *sollemnytee*.

2745 *used*. So MS, C. A: *usid*. R omits.

2746 *arte*. So C, A, R, D&A. MS: fully legible only under UV light.

2753 *is*. So MS, C, A. R: *it is*.

2763 *His bright arisith*. MS: written over an erasure in a darker ink. C, A: *his bright arisich*. R: *his bright arise*. D&A: *his bright aristh*.

mery. So MS, A. C: ~~*mey*~~ *mery*. R: *meri*.

2766 *or ellis snowe*. So C, A, D&A. R: *or els snow*. MS: *or ellisnowe*.

2768 *ayer*. So MS, C, A. R: *rayne*.

2787 *been*. So MS, C, A. R omits.

2790 *wil*. So MS, C, A. R omits.

2822 *ye*. So MS, C, A. R: *the*.

2827 *on*. So MS, R. C: *ion*(?). A: *ion* (*i* erased).

2829 *a*. So MS, C, A. R omits.

2836 *Ye*. So MS, C, A. R omits.

2844 *ye*. So MS, C, A. R: *the*.

2856 *other*. So MS, C, A. R: *over*.

2861 *the*$_2$. So MS. C: *te*. A: *to*. R omits.

2878 *undir*. So MS, A. C: *under*. R: *in the*.

2880 *Harry*. So MS, D&A. C, A: *Edward*. R: *Edwarde*.

2894 *it is*. So MS, C, A. R: *is it*.

2898 *bleede*. So MS, C, A. R: ~~*brel*~~ *bleed*.

2901 *Mischefe*. So MS. C: *Myschief*. A: *Mischeef*. R: *Myscheef*. D&A: *Myschefe*.

2902 *betrapped*. So MS, C. A: *betrappid*. R: *betrayed*.

2909 *Where*. So MS. C, A, R: *Her*.

renoun. So MS, A. C: ~~*re*~~ *renoun*. R: *renoune*.

2910 *forseyn*. So MS. C: *forseyne*. A: *forsaine*. R: *for fame*.

2912 *forwith*. So MS, C. A: *forthwith*. R: *furthwith*.

2913 *Sleuth*. So MS. C, A: *Sleught*. R: *Slouth*.

2915 *Although*. So MS, A. C: *As though*. R: *Allthough*.

2916 *th'*. So MS, C, A. R omits.

2925 *to make*. So MS, R. C omits. A: inserted above the line.

2927 *That*. So C, A, R. MS: first letter written over an erasure.

2930 *either*. So MS, C, A. R: *ever*.

2938 *kene is*. So MS. C, A, R: *is kene*.

2956 *forlete*. So MS, C, A. R: *forto let*.

2958 *to strynge*. So MS, C. A: *to stringe*. R: ~~*to st*~~ *to stryng*.

2960 *with*$_2$. So MS, C, A. R omits.

2962 *her*. So C, A, R, D&A. MS: *h* . . . partially legible only under UV light.

2963 *yon*. So C, A. MS: . . . *on* partially legible only under UV light. R: *yow*. D&A: *[yon]*.

2967 *were*. So C, D&A. MS: *w* . . . *e* partially legible only under UV light. A: *wer*. R: *wher*.

2968 *then*. So C, A, R, D&A. MS: *the*.

2969 *nerf . . . stringis*. So R. MS: fully legible only under UV light. C: *nerf . . . stryngys*. A: *nerf . . . strengis*. D&A: *nerf . . . stringes*.

2970 *made herte*. So C, A, D&A. MS: fully legible only under UV light. R: *mad hert*.

2975 *is*. So MS, A, R. C: *to is*.

2976 *ye*. So MS, C, A. R: *the*.

a point. So MS, A. C: *a poynt*. R omits.

2977 *myn*. So MS, C, A. R: changed to *mayd*.

2978 *me*. So MS, C, A. R omits.

2979 *am I*. So MS, C, A. R: *I am*.

2981 *faynest*. So MS. C, A, R: *fayne*.

londe. So MS. C, A: *o londe*. R: *on land*.

2982 *Joon*. MS, C, A, R: *Ioo*. D&A: *I [—] oo*.

2983 *wrong*. So MS, D&A. C: *werk*. A, R: *werke*.

2984 *wil I not to*. So MS, D&A. C, A: *dar to right go*. R: *dare to right go*.

2985 *litil heer to*. So MS, D&A. C, A, R: *lite in errour*.

2988 *that ayer*. So MS, D&A. C: *ayer*. A: *this ayer* (*this* inserted above the line). R: *this ner*.

2991 *way*. So C, A, D&A. MS: fully legible only under UV light. R: *wey*.

2992 *th'electioun*. So C, A, R, D&A. MS: fully legible only under UV light.

3018 *hemself*. So C, A. MS: fully legible only under UV light. R: *hemselfe*. D&A: *hemself*.

3021 *werk*. So MS, C. A: *werke*. R: *ware*.

3024 *impropur speche*. So MS, C. A: *impropur spech*. R: *inpropure ~~spes~~ speche*. D&A: *impropir speche*.

3025 *amende*. So MS. C, A, R: *tamende*.

3026 *to*. So MS, C, A. R omits.

3027 *ortographie*. So MS, C, A. R: *~~ortogar~~ ortographie*.

after 3028 *Finis*. So MS. C, A, R omit.

Appendix: Concordance to Latin Vegetius

To aid the reader wishing to compare the text of *Knyghthode and Bataile* with the original Latin text of Vegetius's *De re militari* (or any of its other vernacular translations or paraphrases), we have noted the general section number in Vegetius, where applicable, into our editorial subheadings within the edition that follows. As a reference text for Vegetius (and thus of the section numbers and sentence breakdowns in what follows), we have utilized Reeve's edition of the Latin text, not the widely read English translation by Milner. A more specific concordance is provided here, where the line numbers of *Knyghthode and Bataile* (left of each column) are given their corresponding section numbers in Vegetius (right). Indented and italicized line numbers break down correspondences within a larger sction of the text. In what follows:

- ~ indicates a loose paraphrase
- — indicates there is no correspondence to the line(s) within Vegetius

Proem	
1–88	—
Part I	
89–92	1.prol. first sentence only
91	—
93–165	—
166–72	~ Synopsis
173–82	1.2
177	—
183–200	1.3
201–21	1.4
222–28	1.5
229–51	1.6
252–91	1.7
292–319	1.8
320–40	1.9
341–47	1.10
348–75	1.11
349–52	—
376–96	1.12
397–417	1.13
418–31	1.14
432–52	1.15
442–43	—
447	—
453–66	1.16
467–73	1.17
474–80	1.18
481–87	1.19
488–93	2.14
494	1.20 final sentence only
495–501	1.21
502–15	1.22
516–29	1.23
530–50	1.24
551–64	1.25
565–92	1.26
593–613	1.27
614–20	— (1.28 omitted)
Part II	
621–41	— (2.prol. omitted)

624–25	2.4 penultimate sentence only
642–69	2.1
656	—
670–76	2.2 first four sentences only
677–79	2.4 first sentence only (2.3 omitted)
680–705	2.5
706–53	2.6
754–802	2.7
803–51	2.8
852–879	2.9
880–93	2.10
894–907	2.11
908–21	2.12
922–35	2.13
936–77	2.14 (2.15–2.25 omitted)
Part III	
978–1026	— (3.prol. omitted)
1027–82	3.1
1048–52	2.2
1083–1110	3.2
1111–73	3.3
1174–1250	3.4
1179–80	—
1203–07	—
1251–1306	3.5
1276–78	—
1307–1446	3.6
1439	—
1442	—
1447–81	3.7
1466–67	—
1482–1579	3.8
1486–88	—
1499–1500	—
1576	—
1580–1619	3.9
1586	—
1593	—
1598	—
1616	—
1620–21	—
1622–1705	3.10
1656	—
1668–70	—
1681–84	—
1692	—
1706–32	3.11
1733	—
1734–54	3.12
1752	—
1755–61	3.13
1762–1864	3.14
1787	—
1820	—
1849–52	—
1859	—
1865–66	—
1867–1901	3.15
1887	—
1902–22	3.16
1923–51	3.17
1938–39	—
1944–50	—
1952–99	3.18
1967	—
1996	—
2000–06	3.19
2007–45	—
2046–2132	3.20
2076	—
2078	—

2133–60	~ 3.21, 3.22, 3.25 (3.23–3.24 omitted)
2158	—
2161–2216	3.26
2164–67	—
2217–2223	—
Part IV	
2224–44	— (4.prol. omitted)
2245–58	4.1
2259–65	4.2
2266–79	4.3
2280–86	4.4
2287–93	4.5
2294–99	4.6
2300	—
2301–14	4.7
2315–35	4.8
2336–45	4.9
2346–56	4.10
2357–63	4.11
2364–70	4.12
2371–77	4.13
2378–2401	4.14
2402–26	4.23
2427–47	4.15
2436–40	—
2443	—
2448–56	4.16
2449	—
2457–82	4.17
2458	—
2468	—
2471	—
2475	—
2478	—
2483–2503	4.18
2504–10	4.19
2511–20	4.20

2513	—
2521–24	—
2525–38	4.21
2539–41	4.22 first sentence only
2542–52	—
2553–87	4.24, 4.25
2564	—
2571–73	—
2588–2601	4.26
2602–07	4.28 (4.27 omitted)
2608	4.27, 4.28 referred to
2609–11	4.30/4.31 interchapter synopsis
2612–15	4.31 first sentence only
2616–29	4.34 (4.32, 4.33 omitted)
2625	—
2630–50	4.35, 4.36
2651–64	4.37
2665–2706	4.38
2671	—
2691–92	—
2707–48	4.39
2749–73	4.41 (4.40 omitted)
2753	—
2767	—
2774–88	4.42
2789–90	—
2791–2811	4.43
2793–94	—
2800–01	—
2812–34	4.45
2835–2909	~ 4.44, 4.46
2910–27	—
2928–51	~ 4.46
2935	—

2937	—
2941	—
2946–48	—
2952–54	~ 4.22
2955–62	~ 4.9
2963–65	—
2966–72	~ 2.25, 3.24, 4.22
2973–3028	—

Bibliography

Manuscripts and Documents

Cambridge, Pembroke College, MS 243, fols. 1r–55v.
Chippenham, Wiltshire and Swindon Archives, 212B/4796.
Exeter, Devon Record Office, 3248A-0/11/117.
Kew, The National Archives, SC 1/46/265.
London, British Library, Cotton MS Titus A.xxiii, fols. 2r–53v.
London, British Library, Royal MS 18.A.xii.
London, College of Arms, MS R.25, fols. 24r–62v.
Oxford, Bodleian Library, MS Ashmole 45 (Part 2), fols. 1r–7v, 18r–23v, 41r–43v, 46r–53v.
Sankt-Peterburg, Rossiyskaya natsional'naya biblioteka, MS Fr.f.v.IX 1, fols. 1r–58v.
Wolfenbüttel, Herzog August Bibliothek, MS Blankenburg 111, fols. 1r–47r

Primary Sources

Aquinas, Thomas. *Catena Aurea: Commentary on the Four Gospels, Collected out of the Works of the Fathers by S. Thomas Aquinas.* 3 vols. Trans. Mark Pattison, J. D. Dalgrins, and T. D. Ryder. Oxford: John Henry Parker, 1841–45.

———. *The Summa Theologiae.* Trans. Fathers of the English Dominican Province. New York: Benziger Bros., 1947–48. Online at https://aquinas101.thomisticinstitute.org/st-index.

Bartholomaeus Anglicus. *De proprietatibus rerum.* Vol. 1. Ed. Baudouin Van den Abeele, Heinz Meyer, Michael W. Twomey, Bernd Roling, and R. James Long. Turnhout: Brepols, 2007.

Calendar of the Patent Rolls, Preserved in the Public Record Office: Henry VI. 6 vols. London: H. M. Stationery Office, 1901–10.

Carley, Lionel Kenneth, ed. "The Anglo-Norman Vegetius: A Thirteenth-Century Translation of the *De re militari* of Flavius Vegetius Renatus." 2 vols. Ph.D. Dissertation: University of Nottingham, 1962.

Cato (Dionysius Cato). *Distichs of Cato: A Famous Medieval Textbook.* Ed. and trans. Wayland Johnson Chase. Madison: University of Wisconsin Press, 1922.

Chaucer, Geoffrey. *The Riverside Chaucer.* Third edition. Gen. ed. Larry D. Benson. Boston: Houghton Mifflin, 1987.

Clanvowe, John. *The Two Ways.* In *The Works of Sir John Clanvowe.* Ed. V. J. Scattergood. Cambridge: D. S. Brewer, 1975. Pp. 57–80, 86–89.

Dyboski, R., and Z. M. Arend, eds. *Knyghthode and Bataile: A XVth Century Verse Paraphrase of Flavius Vegetius Renatus' Treatise "De re militari."* EETS o.s. 201. London: Oxford University Press, 1935.

Gairdner, James, ed. *The Paston Letters, A. D. 1422–1509.* 6 vols. London: Chatto & Windus, 1904.

Geoffrey of Monmouth. *The History of the Kings of Britain: An Edition and Translation of the "De gestis Britonum" [Historia Regum Britanniae]*. Ed. Michael D. Reeve. Trans. Neil Wright. Woodbridge: Boydell, 2007.

Geoffroi de Charny. *The Book of Chivalry of Geoffroi de Charny: Text, Context, and Translation.* Ed. and trans. Elspeth Kennedy. Introduction by Richard W. Kaeuper. Philadelphia: University of Pennsylvania Press, 1996.

Gower, John. *Confessio Amantis*. Ed. Russell A. Peck, with Latin translations by Andrew Galloway. 3 vols. Kalamazoo, MI: Medieval Institute Publications, 2000–04.

The Holy Bible: Douay-Rheims Version. Ed. Bishop Richard Challoner. Rockford, IL: Tan Books and Publishers, 1989. Originally published Baltimore, MD: John Murphy, 1899. Online at http://www.drbo.org/drl/index.htm.

Homer. *The Iliad*. Trans. Caroline Alexander. New York: Ecco, 2015.

Jean de Meun. *Li abregemenz noble honme Vegesce Flave René des establissemenz apartenanz a chevalerie*. Ed. Leena Löfstedt. Helsinki: Suomalainen Tiedeakatemia, 1977.

Jean de Vignay. *Li livres Flave Végèce de la chose de chevalerie*. Ed. Leena Löfstedt. Helsinki: Suomalainen Tiedeakatemia, 1982.

Jean le Bel. *The True Chronicles of Jean le Bel, 1290–1360*. Trans. Nigel Bryant. Woodbridge: The Boydell Press, 2011.

Jean Priorat. *Li Abrejance de l'Ordre de Chevalerie*. Ed. Ulysse Robert. Paris: Firmin Didot, 1897.

John of Beaumont. *The Vows of the Heron (Les Voeux du héron): A Middle French Vowing Poem*. Ed. John L. Grigsby and Norris J. Lacy. Trans. Norris J. Lacy. New York: Garland, 1992.

Lester, Geoffrey, ed. *The Earliest English Translation of Vegetius' "De re militari."* Heidelberg: Carl Winter Universitätsverlag, 1988.

Löfstedt, Leena, Outi Merisalo, Elina Suomela-Härmä, Renja Salminen, and Lauri Juhani Eerikäinen, eds. *Le livre de l'art de chevalerie de Vegesce: Traduction anonyme de 1380*. Helsinki: Suomalainen Tiedeakatemia, 1989.

Minot, Laurence. *The Poems of Laurence Minot, 1333–1352*. Ed. Richard H. Osberg. Kalamazoo, MI: Medieval Institute Publications, 1996.

Morris, Richard, ed. *A Bestiary*. In *An Old English Miscellany, Containing a Bestiary, Kentish Sermons, Proverbs of Alfred, Religious Poems of the Thirteenth Century, from Manuscripts in the British Museum, Bodleian Library, Jesus College Library, etc.* EETS o.s. 49. Oxford: Oxford University Press, 1872. Rpt. 1927. Pp. 1–25.

Myers, A. R., ed. *English Historical Documents, 1327–1485*. Vol. 4. New York: Oxford University Press, 1969.

Nicolas, Nicholas Harris, ed. *Proceedings and Ordinances of the Privy Council of England*. 7 vols. London: Eyre and Spottiswoode, 1834–37.

Robbins, Rossell Hope, ed. "Defend Us from All Lollardry." In *Historical Poems of the XIVth and XVth Centuries*. New York: Columbia University Press, 1959. Pp. 152–57.

Sallust (Gaius Sallustius Crispus). "The Jugurthine War." In *Catiline's Conspiracy, The Jugurthine War, Histories*. Trans. William W. Batstone. Oxford: Oxford University Press, 2010. Pp. 49–127.

Shakespeare, William. *The Third Part of Henry the Sixth*. In *The Riverside Shakespeare*. Second edition. Ed. G. Blakemore Evans, J. J. M. Tobin, and Herschel Baker. Boston: Houghlin Mifflin, 1997. Pp. 711–47.

Smith, Trevor Russell, trans. "Battle of Winchelsea." In *Medieval Warfare: A Reader*. Ed. Kelly DeVries and Michael Livingston. Toronto: University of Toronto Press, 2019. Pp. 191–93.

Taylor, Frank, and John S. Roskell, eds. and trans. *Gesta Henrici Quinti: The Deeds of Henry the Fifth.* Oxford: Clarendon Press, 1975.

Trevisa, John. *On the Properties of Things: John Trevisa's Translation of Bartholomaeus Anglicus, "De Proprietatibus Rerum": A Critical Text.* Ed. M. C. Seymour. 3 vols. Oxford: Clarendon Press, 1975–88.

Vegetius (Publius Flavius Vegetius Renatus). *Epitome of Military Science.* Trans. N. P. Milner. Second edition. Liverpool: Liverpool University Press, 1996.

———. *Epitoma rei militaris.* Ed. M. D. Reeve. Oxford: Clarendon Press, 2004.

Virgil. *Georgics.* In *Eclogues, Georgics, Aeneid: Books 1–6.* Trans. H. Rushton Fairclough; rev. G. P. Goold. Cambridge, MA: Harvard University Press, 1999.

SECONDARY SOURCES

Abels, Richard, and Stephen Morillo. "A Lying Legacy? A Preliminary Discussion of Images of Antiquity and Altered Reality in Medieval Military History." *Journal of Medieval Military History* 3 (2005), 1–13.

Allmand, Christopher T. "Changing Views of the Soldier in Late Medieval France." In *Guerre et société en France, en Angleterre et en Bourgogne, XIVe–XVe siècle.* Ed. Philippe Contamine, Charles Giry-Deloison, and Maurice H. Keen. Villeneuve d'Ascq: Centre d'histoire de la région du nord et de l'Europe du nord-ouest, Université Charles de Gaulle-Lille III, 1991. Pp. 171–88.

———. "The Fifteenth-Century English Versions of Vegetius' *De Re Militari.*" In *Armies, Chivalry and Warfare in Medieval Britain and France: Proceedings of the 1995 Harlaxton Symposium.* Ed. Matthew Strickland. Stamford: Paul Watkins, 1998. Pp. 30–45.

———. *The "De Re Militari" of Vegetius: The Reception, Transmission and Legacy of a Roman Text in the Middle Ages.* Cambridge: Cambridge University Press, 2011.

———. "The English Translations of Vegetius' *De Re Militari.* What Were their Authors' Intentions?" *The Fifteenth Century XI: Concerns and Preoccupations.* Ed. Linda Clark. Woodbridge: The Boydell Press, 2012. Pp. 1–8.

Anglo, Sydney. "Vegetius's *De re militari*: The Triumph of Mediocrity." *Antiquaries Journal* 82 (2002), 247–67.

Armstrong, C. A. J. "Politics and the Battle of St. Albans, 1455." *Bulletin of the Institute of Historical Research* 33 (1960), 1–72.

Bachrach, Bernard S. "The Practical Use of Vegetius' *De Re Militari* During the Early Middle Ages." *The Historian* 47 (1985), 239–55.

Baker, Gary. "Investigating the Socio-Economic Origins of English Archers in the Second Half of the Fourteenth Century." *Journal of Medieval Military History* 12 (2014), 173–216.

Bell, Adrian R., Anne Curry, Andy King, and David Simpkin. *The Soldier in Later Medieval England.* Oxford: Oxford University Press, 2013.

Boardman, A. W. *The Medieval Soldier in the Wars of the Roses.* Stroud: Sutton, 1998.

Boffey, Julia. "Books and Readers in Calais: Some Notes." *The Ricardian* 13 (2003), 67–74.

Boffey, Julia, and A. S. G. Edwards, eds. *NIMEV: A New Index of Middle English Verse.* London: The British Library, 2005.

Boulton, D'Arcy Jonathan Dacre. *The Knights of the Crown: The Monarchical Orders of Knighthood in Later Medieval Europe, 1325–1520.* Woodbridge: The Boydell Press, 1987.

Charles, Michael B. *Vegetius in Context: Establishing the Date of the "Epitoma Rei Militaris."* Stuttgart: Franz Steiner Verlag, 2007.

Coleman, Arthur Prudden. "Roman Dyboski (19 Nov. 1883–1 June 1945)." *American Slavic and East European Review* 5 (1946), 222–25.

Contamine, Philippe. "Les traités de guerre, de chasse, de blason et de chevalerie." In *La littérature française aux XIVe et XVe siècles*. Vol. 1. Ed. Daniel Poirion. Heidelberg: Carl Winter Universitätsverlag, 1988. Pp. 346–67.

———. "La musique militaire dans le fonctionnement des armées: L'exemple français (v. 1330–v. 1550)." In *Von Crécy bis Mohács: Kriegswesen im späten Mittelalter (1346–1526)*. Wien: Heeresgeschichtliches Museum, 1997. Pp. 93–106.

Curry, Anne. "The Military Ordinances of Henry V: Texts and Contexts." In *War, Government and Aristocracy in the British Isles, c. 1150–1500: Essays in Honour of Michael Prestwich*. Ed. Chris Given-Wilson, Ann Kettle, and Len Scales. Woodbridge: The Boydell Press, 2008. Pp. 214–49.

Davies, Daniel, and A. S. G. Edwards. "A New Manuscript of *Knyghthode and Bataile*." *Medium Ævum* 87 (2018), 137–41.

Dušanić, Slobodan, and Žarko Petković. "The Five Standards of the Pre-Marian Legion. A Note on the Early Plebian *militaria*." *Klio* 85 (2003), 42–56.

Galderisi, Claudio, ed. *Translations médiévales: Cinq siècles de traductions en français au Moyen Âge (XIe–XVe siècles)*. 2 vols. Turnhout: Brepols, 2011.

Goffart, Walter. "The Date and Purpose of Vegetius' *De re militari*." *Traditio* 33 (1977), 65–100.

Jaquet, Daniel, Alice Bonnefoy Mazure, Stéphanie Armand, Caecilia Charbonnier, Jean-Luc Ziltener, and Bengt Kayser. "Range of Motion and Energy Cost of Locomotion of the Late Medieval Armoured Fighter: A Proof of Concept of Confronting the Medieval Technical Literature with Modern Movement Analysis." *Historical Methods* 49 (2016), 169–86.

Jones, Michael K. "Edward IV, the Earl of Warwick and the Yorkist Claim to the Throne." *Historical Research* 70 (1997), 342–52.

Kaeuper, Richard W. *Holy Warriors: The Religious Ideology of Chivalry*. Philadelphia: University of Pennsylvania Press, 2009.

Keegan, John. *The Face of Battle*. New York: Viking Press, 1976.

Keen, Maurice H. *The Laws of War in the Late Middle Ages*. London: Routledge & Kegan Paul, 1965.

Livingston, Michael. "Losses Uncountable: The Context of Crécy." In *The Battle of Crécy: A Casebook*. Pp. 1–13.

Livingston, Michael, and Kelly DeVries, eds. *The Battle of Crécy: A Casebook*. Liverpool: Liverpool University Press, 2015.

MacCracken, Henry Noble. "Vegetius in English: Notes on the Early Translations." In *Anniversary Papers by Colleagues and Pupils of George Lyman Kittredge*. Boston: Ginn and Co., 1913. Pp. 389–403.

Marvin, Julia. *The Construction of Vernacular History in the Anglo-Norman Prose "Brut" Chronicle: The Manuscript Culture of Late Medieval England*. Woodbridge: York Medieval Press, 2017.

Matheson, Lister M. *The Prose "Brut": The Development of a Middle English Chronicle*. Tempe, AZ: Medieval & Renaissance Texts & Studies, 1998.

Meyer, Heinz. *Die Enzyklopädie des Bartholomäus Anglicus. Untersuchungen zur Überlieferungs- und Rezeptionsgeschichte von "De proprietatibus rerum."* München: Wilhelm Fink, 2000.

Middle English Dictionary. Ann Arbor: University of Michigan Press, 2001–. Online at http://quod.lib.umich.edu/m/med/.

Nall, Catherine. "The Production and Reception of Military Texts in the Aftermath of the Hundred Years War." Ph.D. Dissertation: University of York, 2004.

———. "Perceptions of Financial Mismanagement and the English Diagnosis of Defeat." In *The Fifteenth Century VII: Conflicts, Consequences and the Crown in the Late Middle Ages*. Ed. Linda Clark. Woodbridge: The Boydell Press, 2007. Pp. 119–35.

———. *Reading and War in Fifteenth-Century England: From Lydgate to Malory*. Cambridge: D. S. Brewer, 2012.

Norri, Juhani, ed. *Dictionary of Medical Vocabulary in English, 1375–1550*. 2 vols. New York: Ashgate, 2016.

The Oxford English Dictionary. Third edition. Oxford University Press, 2018. Online at http://www.oed.com/.

Peck, Russell A. *Kingship and Common Profit in Gower's Confessio Amantis*. Carbondale: Southern Illinois University Press, 1978.

Prestwich, Michael. *Armies and Warfare in the Middle Ages: The English Experience*. New Haven: Yale University Press, 1996.

Reeve, Michael D. "The Transmission of Vegetius's *Epitoma rei militaris*." *Aevum* 74 (2000), 243–354.

Rogers, Clifford J. "Gunpowder Artillery in Europe, 1326–1500: Innovation and Impact." In *Technology, Violence, and War: Essays in Honor of Dr John F. Guilmartin, Jr*. Ed. Robert S. Ehlers, Jr., Sarah K. Douglas, and Daniel P. M. Curzon. Leiden: Brill, 2019. Pp. 39–71.

Scase, Wendy. "Writing and the 'Poetics of Spectacle': Political Epiphanies in *The Arrivall of Edward IV* and Some Contemporary Lancastrian and Yorkist Texts." In *Images, Idolatry, and Iconoclasm in Late Medieval England: Textuality and the Visual Image*. Ed. Jeremy Dimmick, James Simpson, and Nicolette Zeeman. Oxford: Oxford University Press, 2002. Pp. 172–84.

Scattergood, John. "The Date of Sir John Clanvowe's *The Two Ways* and the 'Reinvention of Lollardy'." *Medium Ævum* 79 (2010), 116–20.

Sherwood, Foster Hallberg. "Studies in Medieval Uses of Vegetius' *Epitoma rei militaris*." Ph.D. Dissertation: University of California at Los Angeles, 1980.

Smith, Robert Douglas, and Kelly DeVries. *The Artillery of the Dukes of Burgundy, 1363–1477*. Woodbridge: The Boydell Press, 2005.

Smith, Trevor Russell. "National Identity, Propaganda, and the Ethics of War in English Historical Literature, 1327–77." Ph.D. Dissertation: University of Leeds, 2017.

Strickland, Matthew. *War and Chivalry: The Conduct and Perception of War in England and Normandy, 1066–1217*. Cambridge: Cambridge University Press, 1996.

Taylor, Andrew. "Chivalric Conversation and the Denial of Male Fear." In *Conflicted Identities and Multiple Masculinities: Men in the Medieval West*. Ed. Jacqueline Murray. New York: Garland, 1999. Pp. 169–88.

Taylor, Craig. "Military Courage and Fear in the Late Medieval French Chivalric Imagination." *Cahiers de recherches médiévales et humanistes* 24 (2012), 129–47.

———. *Chivalry and the Ideals of Knighthood in France during the Hundred Years War*. Cambridge: Cambridge University Press, 2013.

———. "Henry V, Flower of Chivalry." In *Henry V: New Interpretations*. Ed. Gwilym Dodd. Woodbridge: York Medieval Press, 2013. Pp. 217–47.

Vale, Malcolm. *War and Chivalry: Warfare and Aristocratic Culture in England, France and Burgundy at the End of the Middle Ages*. Athens: University of Georgia Press, 1981.

Wakelin, Daniel. "The Occasion, Author, and Readers of *Knyghthode and Bataile*." *Medium Ævum* 73 (2004), 260–72.

———. "Scholarly Scribes and the Creation of *Knyghthode and Bataile*." *English Manuscripts Studies 1100–1700* 12 (2005), 26–45.

———. *Humanism, Reading, and English Literature, 1430–1530*. Oxford: Oxford University Press, 2007.

Whetham, David. *Just Wars and Moral Victories: Surprise, Deception and the Normative Framework of European War in the Later Middle Ages*. Leiden: Brill, 2009.

Whiting, Bartlett Jere, and Helen Prescott Whiting. *Proverbs, Sentences, and Proverbial Phrases from English Writings Mainly before 1500*. Cambridge, MA: Belknap, 1968.

Middle English Texts Series

The Floure and the Leafe, The Assembly of Ladies, The Isle of Ladies, edited by Derek Pearsall (1990)
Three Middle English Charlemagne Romances, edited by Alan Lupack (1990)
Six Ecclesiastical Satires, edited by James M. Dean (1991)
Heroic Women from the Old Testament in Middle English Verse, edited by Russell A. Peck (1991)
The Canterbury Tales: Fifteenth-Century Continuations and Additions, edited by John M. Bowers (1992)
Gavin Douglas, *The Palis of Honoure*, edited by David Parkinson (1992)
Wynnere and Wastoure and The Parlement of the Thre Ages, edited by Warren Ginsberg (1992)
The Shewings of Julian of Norwich, edited by Georgia Ronan Crampton (1994)
King Arthur's Death: The Middle English Stanzaic Morte Arthur and Alliterative Morte Arthure, edited by Larry D. Benson, revised by Edward E. Foster (1994)
Lancelot of the Laik and Sir Tristrem, edited by Alan Lupack (1994)
Sir Gawain: Eleven Romances and Tales, edited by Thomas Hahn (1995)
The Middle English Breton Lays, edited by Anne Laskaya and Eve Salisbury (1995)
Sir Perceval of Galles and Ywain and Gawain, edited by Mary Flowers Braswell (1995)
Four Middle English Romances: Sir Isumbras, Octavian, Sir Eglamour of Artois, Sir Tryamour, edited by Harriet Hudson (1996; second edition 2006)
The Poems of Laurence Minot, 1333–1352, edited by Richard H. Osberg (1996)
Medieval English Political Writings, edited by James M. Dean (1996)
The Book of Margery Kempe, edited by Lynn Staley (1996)
Amis and Amiloun, Robert of Cisyle, and Sir Amadace, edited by Edward E. Foster (1997; second edition 2007)
The Cloud of Unknowing, edited by Patrick J. Gallacher (1997)
Robin Hood and Other Outlaw Tales, edited by Stephen Knight and Thomas Ohlgren (1997; second edition 2000)
The Poems of Robert Henryson, edited by Robert L. Kindrick with the assistance of Kristie A. Bixby (1997)
Moral Love Songs and Laments, edited by Susanna Greer Fein (1998)
John Lydgate, *Troy Book Selections*, edited by Robert R. Edwards (1998)
Thomas Usk, *The Testament of Love*, edited by R. Allen Shoaf (1998)
Prose Merlin, edited by John Conlee (1998)
Middle English Marian Lyrics, edited by Karen Saupe (1998)
John Metham, *Amoryus and Cleopes*, edited by Stephen F. Page (1999)
Four Romances of England: King Horn, Havelok the Dane, Bevis of Hampton, Athelston, edited by Ronald B. Herzman, Graham Drake, and Eve Salisbury (1999)
The Assembly of Gods: Le Assemble de Dyeus, or Banquet of Gods and Goddesses, with the Discourse of Reason and Sensuality, edited by Jane Chance (1999)
Thomas Hoccleve, *The Regiment of Princes*, edited by Charles R. Blyth (1999)
John Capgrave, *The Life of Saint Katherine*, edited by Karen A. Winstead (1999)
John Gower, *Confessio Amantis*, Vol. 1, edited by Russell A. Peck; with Latin translations by Andrew Galloway (2000; second edition 2006); Vol. 2 (2003); Vol. 3 (2004)
Richard the Redeless and Mum and the Sothsegger, edited by James M. Dean (2000)
Ancrene Wisse, edited by Robert Hasenfratz (2000)
Walter Hilton, *The Scale of Perfection*, edited by Thomas H. Bestul (2000)
John Lydgate, *The Siege of Thebes*, edited by Robert R. Edwards (2001)
Pearl, edited by Sarah Stanbury (2001)

The Trials and Joys of Marriage, edited by Eve Salisbury (2002)
Middle English Legends of Women Saints, edited by Sherry L. Reames, with the assistance of Martha G. Blalock and Wendy R. Larson (2003)
The Wallace: Selections, edited by Anne McKim (2003)
Richard Maidstone, *Concordia (The Reconciliation of Richard II with London)*, edited by David R. Carlson, with a verse translation by A. G. Rigg (2003)
Three Purgatory Poems: The Gast of Gy, Sir Owain, The Vision of Tundale, edited by Edward E. Foster (2004)
William Dunbar, *The Complete Works*, edited by John Conlee (2004)
Chaucerian Dream Visions and Complaints, edited by Dana M. Symons (2004)
Stanzaic Guy of Warwick, edited by Alison Wiggins (2004)
Saints' Lives in Middle English Collections, edited by E. Gordon Whatley, with Anne B. Thompson and Robert K. Upchurch (2004)
Siege of Jerusalem, edited by Michael Livingston (2004)
The Kingis Quair and Other Prison Poems, edited by Linne R. Mooney and Mary-Jo Arn (2005)
The Chaucerian Apocrypha: A Selection, edited by Kathleen Forni (2005)
John Gower, *The Minor Latin Works*, edited and translated by R. F. Yeager, with *In Praise of Peace*, edited by Michael Livingston (2005)
Sentimental and Humorous Romances: Floris and Blancheflour, Sir Degrevant, The Squire of Low Degree, The Tournament of Tottenham, and The Feast of Tottenham, edited by Erik Kooper (2006)
The Dicts and Sayings of the Philosophers, edited by John William Sutton (2006)
"Everyman" and Its Dutch Original, "Elckerlijc," edited by Clifford Davidson, Martin W. Walsh, and Ton J. Broos (2007)
The N-Town Plays, edited by Douglas Sugano, with assistance by Victor I. Scherb (2007)
The Book of John Mandeville, edited by Tamarah Kohanski and C. David Benson (2007)
John Lydgate, *The Temple of Glas*, edited by J. Allan Mitchell (2007)
The Northern Homily Cycle, edited by Anne B. Thompson (2008)
Codex Ashmole 61: A Compilation of Popular Middle English Verse, edited by George Shuffelton (2008)
Chaucer and the Poems of "Ch," edited by James I. Wimsatt (revised edition 2009)
William Caxton, *The Game and Playe of the Chesse*, edited by Jenny Adams (2009)
John the Blind Audelay, *Poems and Carols*, edited by Susanna Fein (2009)
Two Moral Interludes: The Pride of Life and Wisdom, edited by David Klausner (2009)
John Lydgate, *Mummings and Entertainments*, edited by Claire Sponsler (2010)
Mankind, edited by Kathleen M. Ashley and Gerard NeCastro (2010)
The Castle of Perseverance, edited by David N. Klausner (2010)
Robert Henryson, *The Complete Works*, edited by David J. Parkinson (2010)
John Gower, *The French Balades*, edited and translated by R. F. Yeager (2011)
The Middle English Metrical Paraphrase of the Old Testament, edited by Michael Livingston (2011)
The *York Corpus Christi Plays*, edited by Clifford Davidson (2011)
Prik of Conscience, edited by James H. Morey (2012)
The Dialogue of Solomon and Marcolf: A Dual-Language Edition from Latin and Middle English Printed Editions, edited by Nancy Mason Bradbury and Scott Bradbury (2012)
Croxton Play of the Sacrament, edited by John T. Sebastian (2012)
Ten Bourdes, edited by Melissa M. Furrow (2013)

Lybeaus Desconus, edited by Eve Salisbury and James Weldon (2013)

The Complete Harley 2253 Manuscript, Vol. 2, edited and translated by Susanna Fein with David Raybin and Jan Ziolkowski (2014); Vol. 3 (2015); Vol. 1 (2015)

Oton de Granson, Poems, edited and translated by Peter Nicholson and Joan Grenier-Winther (2015)

The King of Tars, edited by John H. Chandler (2015)

John Hardyng Chronicle, edited by James Simpson and Sarah Peverley (2015)

Richard Coer de Lyon, edited by Peter Larkin (2015)

Guillaume de Machaut, The Complete Poetry and Music, Volume 1: The Debate Poems, edited and translated by R. Barton Palmer (2016)

Lydgate's Fabula Duorum Mercatorum and Guy of Warwyk, edited by Pamela Farvolden (2016)

The Katherine Group (MS Bodley 34), edited by Emily Rebekah Huber and Elizabeth Robertson (2016)

Sir Torrent of Portingale, edited by James Wade (2017)

The Towneley Plays, edited by Garrett P. J. Epp (2018)

The Digby Mary Magdalene Play, edited by Theresa Coletti (2018)

Guillaume de Machaut, The Complete Poetry and Music, Volume 9: The Motets, edited by Jacques Boogart (2018)

Six Scottish Courtly and Chivalric Poems, Including Lyndsay's Squyer Meldrum, edited by Rhiannon Purdie and Emily Wingfield (2018)

Gavin Douglas, *The Palyce of Honour*, edited by David John Parkinson (2018)

Guillaume de Machaut, The Complete Poetry and Music, Volume 2: The Boethian Poems, Le Remede de Fortune and Le Confort d'Ami, edited by R. Barton Palmer (2019)

John Lydgate's "Dance of Death" and Related Works, edited by Megan L. Cook and Elizaveta Strakhov (2019)

The Roland and Otuel Romances and the Anglo-French Otinel, edited by Elizabeth Melick, Susanna Fein, and David Raybin (2020)

Christine de Pizan's Advice for Prices in Middle English Translation: Stephen Scrope's "The Epistle of Othea" and the Anonymous "Lytle Bibell of Knyghthod," edited by Misty Schieberle (2020)

Commentary Series

Haimo of Auxerre, *Commentary on the Book of Jonah*, translated with an introduction and notes by Deborah Everhart (1993)

Medieval Exegesis in Translation: Commentaries on the Book of Ruth, translated with an introduction and notes by Lesley Smith (1996)

Nicholas of Lyra's Apocalypse Commentary, translated with an introduction and notes by Philip D. W. Krey (1997)

Rabbi Ezra Ben Solomon of Gerona, *Commentary on the Song of Songs and Other Kabbalistic Commentaries*, selected, translated, and annotated by Seth Brody (1999)

John Wyclif, *On the Truth of Holy Scripture*, translated with an introduction and notes by Ian Christopher Levy (2001)

Second Thessalonians: Two Early Medieval Apocalyptic Commentaries, introduced and translated by Steven R. Cartwright and Kevin L. Hughes (2001)

The "Glossa Ordinaria" on the Song of Songs, translated with an introduction and notes by Mary Dove (2004)

The Seven Seals of the Apocalypse: Medieval Texts in Translation, translated with an introduction and notes by Francis X. Gumerlock (2009)

The "Glossa Ordinaria" on Romans, translated with an introduction and notes by Michael Scott Woodward (2011)

Nicholas of Lyra, *Literal Commentary on Galatians*, translated with an introduction and notes by Edward Arthur Naumann (2015)

Early Latin Commentaries on the Apocalypse, edited by Francis X. Gumerlock (2016)

Rabbi Eliezer of Beaugency: Commentaries on Amos and Jonah (with selections from Isaiah and Ezekiel), by Robert A. Harris (2018)

Carolingian Commentaries on the Apocalypse by Theodulf and Smaragdus, edited and translated by Francis X. Gumerlock (2019)

Secular Commentary Series

Accessus ad auctores: Medieval Introduction to the Authors, edited and translated by Stephen M. Wheeler (2015)

The Vulgate Commentary on Ovid's Metamorphoses, Book 1, edited and translated by Frank Coulson (2015)

Brunetto Latini, *La rettorica*, edited and translated by Stefania D'Agata D'Ottavi (2016)

Documents of Practice Series

Love and Marriage in Late Medieval London, selected, translated, and introduced by Shannon McSheffrey (1995)

Sources for the History of Medicine in Late Medieval England, selected, introduced, and translated by Carole Rawcliffe (1995)

A Slice of Life: Selected Documents of Medieval English Peasant Experience, edited, translated, and with an introduction by Edwin Brezette DeWindt (1996)

Regular Life: Monastic, Canonical, and Mendicant "Rules," selected and introduced by Douglas J. McMillan and Kathryn Smith Fladenmuller (1997); second edition, selected and introduced by Daniel Marcel La Corte and Douglas J. McMillan (2004)

Women and Monasticism in Medieval Europe: Sisters and Patrons of the Cistercian Reform, selected, translated, and with an introduction by Constance H. Berman (2002)

Medieval Notaries and Their Acts: The 1327–1328 Register of Jean Holanie, introduced, edited, and translated by Kathryn L. Reyerson and Debra A. Salata (2004)

John Stone's Chronicle: Christ Church Priory, Canterbury, 1417–1472, selected, translated, and introduced by Meriel Connor (2010)

Medieval Latin Liturgy in English Translation, edited by by Matthew Cheung Salisbury (2017)

Henry VII's London in the Great Chronicle, edited by Julia Boffey (2019)

Medieval German Texts in Bilingual Editions Series

Sovereignty and Salvation in the Vernacular, 1050–1150, introduction, translations, and notes by James A. Schultz (2000)

Ava's New Testament Narratives: "When the Old Law Passed Away," introduction, translation, and notes by James A. Rushing, Jr. (2003)

History as Literature: German World Chronicles of the Thirteenth Century in Verse, introduction, translation, and notes by R. Graeme Dunphy (2003)

Thomasin von Zirclaria, *Der Welsche Gast (The Italian Guest)*, translated by Marion Gibbs and Winder McConnell (2009)

Ladies, Whores, and Holy Women: A Sourcebook in Courtly, Religious, and Urban Cultures of Late Medieval Germany, introductions, translations, and notes by Ann Marie Rasmussen and Sarah Westphal-Wihl (2010)

Neidhart: Selected Songs from the Riedegg Manuscript, introduction, translation, and commentary by Kathryn Starkey and Edith Wenzel (2016)

Varia

The Study of Chivalry: Resources and Approaches, edited by Howell Chickering and Thomas H. Seiler (1988)

Studies in the Harley Manuscript: The Scribes, Contents, and Social Contexts of British Library MS Harley 2253, edited by Susanna Fein (2000)

The Liturgy of the Medieval Church, edited by Thomas J. Heffernan and E. Ann Matter (2001; second edition 2005)

Johannes de Grocheio, *Ars musice*, edited and translated by Constant J. Mews, John N. Crossley, Catherine Jeffreys, Leigh McKinnon, and Carol J. Williams (2011)

Aribo, "De musica" and "Sententiae," edited and translated by T. J. H. McCarthy (2015)

Guy of Saint-Denis, *Tractatus de tonis*, edited and translated by Constant J. Mews, Carol J. Williams, John N. Crossley, and Catherine Jeffreys (2017)

Typeset in 11/13.2 Arno Pro
and Golden Cockerel Ornaments display

Medieval Institute Publications
College of Arts and Sciences
Western Michigan University
1903 W. Michigan Avenue
Kalamazoo, MI 49008-5432
http://www.wmich.edu/medievalpublications

Western Michigan University